TEACH YOURSELF BOOKS

This book is intended as an introduction to modern electronics for the student and the interested layman. It gives a simple but detailed account of the basic electronic devices and circuits, and discusses their application in communication, computation and control systems. The final chapter describes briefly some of the more recent developments in the field of electronics, including lasers, travelling wave tubes, cryoelectronics and thyristors. The text is illustrated throughout. It is hoped that the reader will gain sufficient knowledge of the fundamental principles and techniques of the subject not to be intimidated by any electronics he subsequently meets.

THE AUTHOR

After training as a physicist Professor W. P. Jolly became a naval radar officer during the 1939–45 war. He is now Visiting Teacher at the Department of Electronic and Electrical Engineering, King's College, University of London. He is a consultant in electronic instrumentation and business application of scientific principles. The author of several books on physics and electronics, Professor Jolly has also written the biographies of Marconi and Oliver Lodge.

TEACH YOURSELF BOOKS

ELECTRONICS

W. P. Jolly,

B.Sc., C.Eng., F.I.E.E., F.I.E.R.E.

ST. PAUL'S HOUSE WARWICK LANE
LONDON EC4P 4AH

First printed 1972
Second edition 1975

ISBN 0 340 19410 3

Printed and bound in Great Britain
for The English Universities Press Ltd.
by Richard Clay (The Chaucer Press) Ltd., Bungay, Suffolk

Contents

List of Plates

Foreword

'You can't see the wood for the trees' is an overworked phrase which has a certain validity but overlooks the fact that it is individual trees which stop the traveller, not the wood.

World vegetation and electronics are subjects of comparable complexity. The professional botanist will have an encyclopaedic knowledge acquired over a long period, but the beginner or the intelligent and inquisitive layman will want to grasp the essentials of the subject quickly. When commencing the study of world vegetation, a high ratio of understanding achieved to time spent might be reached by considering the *differences* between the various types of 'wood'—pine forest, thicket, jungle—and the *similarities* among all trees—roots, branches, leaves. Such an approach has been followed in presenting electronics in this book.

The opening chapter is essentially a self-contained review of modern electronics presenting the fundamental concepts of the electron, energy and waves, and the general principles applied in systems for communication, computation and control. The following chapters set out the basic nature and properties of the materials used in electronics and describe the valve, the junction diode and the transistor so that the reader may understand in detail the working of the basic switching and amplifying circuits. The way in which electronic switches and amplifiers may be organised into circuits for special functions is next described, and these special circuits, treated as blocks, are then shown collected together in typical systems. A final chapter looks at electronic developments outside the mainstream already described so that the reader may feel that he knows something about lasers, travelling wave tubes, cryoelectronics, thyristors, and other devices and techniques that he is certain to hear of in everyday life.

In essence, this book sets out to 'do' modern electronics in about two hundred small pages so that the reader will recognise and understand both the 'woods' and the 'trees'. In particular, it is hoped that he will not be intimidated by any electronics he subsequently encounters, but will feel that it must apply principles he has met before and use circuit elements developed from those whose detailed operation he has followed.

WPJ

Acknowledgment

The author and publishers would like to thank Mullard Limited for their kind permission to reproduce Plates 1 to 4 in this book.

Publishers' Note

American readers of this book should note that what is termed a 'valve' in the U.K. is called a 'tube' in the U.S.A.

CHAPTER 1

Why Electronics?

The question put colloquially in the title of this introductory chapter really needs two answers.

Why choose electronics as a subject for study? Perhaps the quick, sad answer is that there is an examination to be passed. But an examination passed means knowledge acquired. Is electronics important enough for this knowledge to be worth acquiring, even with no examination to pass?

There should be reassurance enough for any reader in the enormous existing range of the electronics industry. Instrumentation, communication and computation involve such colossal capital investment and such comprehensive interaction with daily life that there is little doubt that the subject would already merit study in its own right, even if no further development ever took place.

This judgment takes no account of those new technologies, new arts and new entertainments which have grown from electronics, or which depend upon electronics for their existence. People with interests or work in these fields need not necessarily have a knowledge of the principles of electronics, any more than Michelangelo needed his knowledge of the techniques of quarrying marble. But such technical knowledge is bound to throw light on problems, enlarge horizons and generally increase expertise—and enjoyment—in areas which may be far removed from technology.

Another question could well be: 'Why electronics rather than neutronics or hydrogen atomics?' Is there something special about the electron which distinguishes it from other particles? Perhaps we should at this stage review briefly some of the principal properties of the electron which are responsible for its fundamental importance and for the many ways it has been used in engineering systems.

The electron is the smallest, lightest object which can exist, and it can easily be moved faster than any other object. For these reasons it clearly has great potential in technology.

It is because an electron carries an electric charge that it is particularly easy to manipulate by electrical methods. In fact, the electron *is* electricity: an electric current is a flow of electrons, and electricity, in its uncomplicated role of provider of heat, light and motive power, has a whole set of industries of its own.

We shall see that the electron is even more fundamental, because it is a part of every material, and anything which happens to a substance will be reflected in some sort of rearrangement of its constituent electrons. This rearrangement can usually be detected by some suitable electronic instrument if we want to find out what is happening to the substance. Conversely, if we want to make something happen to the substance, it can often be done electronically.

The substance which most concerns us, even if we are not hypochondriacs, is the material of our own bodies. Electrons are part of every speck and fibre of every living creature, and, flowing in the brain, the nerves and the muscles, they determine our thoughts and movements.

Electrons radiate the heat and light that determine our natural environment, and if this environment is not to our liking, then electricity can be used to change it.

If we want to talk to someone out of earshot, or look

at something out of sight, or do a sum that would take more than a lifetime, then electronics will let us do it.

The electron is the stuff we are made of, the agent of our thoughts and movements, the controller of our environment and the extension of our senses. The study of modern electronics will tell us much more than merely what goes on inside the 'black boxes' which play such a large part in modern living—and dying. It can also tell us something about man, and how he may interact with his environment and with other men.

All this is rather grandiose and it is time we had a look at the specific topics we intend to discuss in order to make contact with the broad generalisations put forward above.

Structure of Matter

One topic which we shall certainly discuss is the structure of matter. We shall find that there is a simple, general pattern in the structure of all the solid materials which principally concern us, and that electrons are a key part of that pattern.

There are ninety-two separate chemical substances from which the world is made; substances such as oxygen, carbon, hydrogen, iron, sulphur and silicon. Often, of course, they are combined together to make more elaborate materials, such as hydrogen and oxygen in water. But these are the ninety-two elements; elements which the chemists have grouped in a list called the Periodic Table and which, in various combinations, go to make up all the millions of compounds that exist.

If, however, an atom of any one of these elements is examined, it will be found to consist of an assembly of three different kinds of particle: protons, electrons and neutrons. The atoms of some elements may contain several hundred particles, while other elements may have less than ten particles in the atom.

Before we look at the way in which these particles make up an atom, we need to know something of their two most important properties: mass and electrical charge. The proton and the neutron both have approximately the same mass, and the mass of the electron is very much less—about $\frac{1}{1840}$ of the mass of the proton. The electron is negatively charged, and the proton has a charge of the same size but of positive sign. The neutron carries no charge.

Any object, from an atom upwards in size, will normally contain equal numbers of protons and electrons, and will thus have no net electrical charge. If electrons are removed in some way from the object, it will be left with a net positive charge. If electrons are added to the object, it will become negatively charged.

Two objects which are electrically charged exert a force on each other which is inversely proportional to the square of their distance apart. If the two charges have the same sign, then the objects repel each other, and if the charges are of opposite sign, then they attract each other. In particular, a proton and an electron will attract each other, and the closer they are together, the greater will be the force.

In general, an atom has a central core called the nucleus, which consists of protons and neutrons. Surrounding this nucleus is a cloud of electrons. The number of protons in the nucleus is equal to the number of electrons in the cloud. The total positive charge on the nucleus due to all the protons is just balanced by the total negative charge of all the electrons in the cloud, and the atom as a whole is electrically neutral.

To get an idea of the principles common to the structure of all atoms we shall start by considering the simplest atom—that of the element hydrogen.

Hydrogen is the lightest of all the atoms, with a single electron in the electron 'cloud' and a single proton as its nucleus. All other elements have neutrons as well as protons in the nucleus; for example, helium, the next simplest atom, has two electrons in the cloud, with two protons and two neutrons in the nucleus.

In the hydrogen atom the electron rotates round the nucleus—in this case a single proton—rather like the Earth rotates round the Sun. If we think of the circular electron orbit, then its radius is such that the electrical attraction between the positively charged proton and the negatively charged electron is just sufficient to provide the force needed to bend the electron path into a circle, just as a stone whirled round on a string would fly off at a tangent if it were not constrained to its circular path by the tension in the string.

This simple system with the electron rotating round the nucleus has a certain amount of energy associated with it.

First, there is the kinetic energy of the moving electron. Any moving body possesses kinetic energy, and the amount of energy is proportional to the mass of the body and the square of its velocity. Thus a small, high-speed body like a bullet may have more kinetic energy than a much larger body moving slowly.

The second kind of energy associated with the hydrogen atom is potential energy, due to the fact that a positive and negative charge separated by a certain distance attract each other, and could be organised to do work in coming together. In the same way, a lake of water at the top of a mountain has potential energy associated with it, because it could be organised to do work by running through turbines down into the valley. The higher the lake is above the valley, the greater will be the potential energy. Similarly, the farther the electron is from the nucleus in the hydrogen atom, the greater will be the potential energy.

The total energy associated with the hydrogen atom is the sum of the potential energy and the kinetic energy. This total depends upon the radius of the orbit in which the electron rotates and has a minimum value for hydrogen atoms in the normal state. If a normal hydrogen atom is in some way given a little extra energy, then the electron moves out into an orbit of greater radius, and the total energy associated with the atom is now greater than it was in the normal state. Atoms possessing more than the normal amount of energy are said to be 'excited'.

The way in which atoms receive extra energy to go into an excited state, and the way in which they give up that energy in returning to the normal state, is of fundamental importance and we shall discuss it later as the quantum theory. In particular, we shall find that energy can only be given to an atom in packets of certain sizes and that energy is emitted by excited atoms in similar packets. The wrong size packet will not be accepted by an atom, and an excited atom will never emit anything other than one of a limited set of packet sizes.

The fact that an atom of any particular element can emit or absorb energy only in packets of certain sizes is one consequence of a set of rules which governs the behaviour of electrons in atoms. These rules also give rise to two other important general properties of the electrons in atoms.

The first of these is that, within the atom, an electron may only possess certain energies—there are certain 'permitted energy levels' for the electron. This concept of permitted energy levels is extended later from the single isolated atom to the electrons in solids, where the electrical properties are largely determined by the permitted energy levels and which of them are possessed by electrons.

Another consequence of the rules for electron behaviour within the atom is one which concerns us immediately, because these rules establish a set of patterns into which the electrons are arranged in more elaborate atoms than those of hydrogen.

The hydrogen atom is very simple, with its single electron rotating in a circular orbit, which is of a fixed radius for the normal state of the atom.

Helium is the next simplest atom, with two electrons and, of course, two protons in the nucleus to balance the negative charge of the electrons. The nucleus also contains two neutrons, and the two electrons circulate round this nucleus in the same orbit.

When we come to the next element, lithium, which has three orbital electrons, the electrons are arranged in a new way round the nucleus. The nucleus now contains three protons and some neutrons.

Two of the electrons are in the same orbit, but the third one is in a different orbit, farther away from the nucleus. The rules say that the innermost orbit, or shell, is completely filled when it has two electrons. Atoms like lithium, with more than two electrons, must start a second orbit of greater radius to accommodate the third, fourth, etc., electrons. When this second orbit has eight electrons in it, there is no room for more, and a third orbit of still larger radius has to be started. Thus sodium, which has eleven orbital electrons, has two in the innermost orbit, eight in the next orbit and one in the outermost orbit.

In the illustrations of electron orbits for various atoms,

the number with the positive sign indicates the number of protons in the nucleus and thus is equal to the number of orbital electrons. This number is called the atomic number of the element. There are neutrons in all the nuclei except hydrogen.

The chemical properties of an element are determined by the electrons, and in this respect it is not surprising

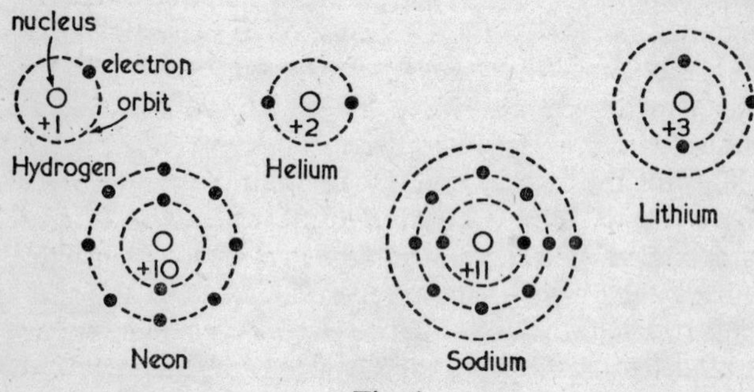

Fig. 1

that the outermost electrons are most important because, when two atoms come together in a chemical reaction, it is the electrons in the outside orbits which will first meet and interact. It is worth noting from the illustrations that hydrogen, lithium and sodium, all with a single electron in their outside shell, have a general similarity of chemical behaviour. Similarly, many of the atoms with two electrons—or with three, etc.—in the outer shell, can be grouped together as being related chemically.

Atoms do not normally have a separate existence, and the simplest form in which we meet matter in the natural world is the gas. This consists of single molecules of the substance moving about at random and largely independent of each other except when they collide.

Typical gas molecules are oxygen (O_2) and hydrogen (H_2). In each of these molecules, two atoms of the element are joined together to make the stable unit found in natural oxygen or hydrogen. The atoms in a molecule are bound together by forces due to complex interactions between the electron systems of the individual atoms. The exact nature of these forces does not concern us as we shall be more interested in solids than in gases. However, it is important to notice that molecules, like atoms, emit or absorb energy in packets of certain sizes and that the electrons in the molecular system also have only certain permitted energies.

Gas molecules may contain atoms of more than one element, e.g. carbon dioxide (CO_2), and in the case of some organic gases may contain many atoms of several different elements and be very large.

The solid is in many ways similar to a very large molecule. Most of the solids that are important in electronics are crystalline, and the characteristic feature of a crystal is a regular arrangement of atoms. The atoms may all be of the same element, as in copper, or they may be of different elements, as in common salt ($NaCl$) or copper sulphate ($CuSO_4$).

In the crystal, as in the molecule, the interaction of the orbital electrons binds the individual atoms into the characteristic pattern or lattice.

There are certain permitted energy levels for electrons in the solid, just as there are in the atom and the molecule, and these energy levels will be different in different materials.

One special aspect of electron properties in solids is that certain materials, particularly metals, contain some electrons which are able to move away from their parent atoms. If a battery is connected to such a material, electrons will flow through the material, and it is called a conductor of electricity. If the application of a battery to a material does not cause a flow of electric current, then the material is called an insulator.

As they flow through a conductor, the electrons which make up the current being driven round the circuit by the battery give up some of their energy to the main solid structure of the material, which thus becomes hot. If electrons lose much energy in passing through a conducting material, then that material is said to have a large resistance to current flow. A given current flowing through a high-resistance conductor will generate much more heat than the same current flowing through a low-resistance conductor. Thus the heating element of an electric fire will be made of high-resistance material, usually a metal alloy. The wires carrying the current under the floor to the fire will be made of low-resistance material, invariably a metal. Materials which offer low resistance to the passage of an electric current are called good electrical conductors.

This, then, is the way in which the chemical energy stored in a battery is converted into heat through the action of a flow of electrons in a conductor. If a metal is heated to a sufficiently high temperature by the current, as in the tungsten lamp filament, then the same mechanism can provide light—another form of energy.

Sometimes, however, it will be necessary to encourage electrons actually to escape from the surface of the metal so that, for instance, they can be formed into a beam passing down a cathode-ray tube to paint a picture on a screen.

Electron Emission

The emission of electrons from a material occurs only if the electrons inside are in some way given sufficient energy to break through the surface where a sort of energy barrier exists, called the work function of the material. Materials with low work functions emit electrons easily because less energy is required for an electron to overcome the energy barrier and escape.

If a solid is heated, then it receives extra energy. This may be shared between the regular crystal lattice and some of the electrons which can escape from their parent atoms. If the electrons thereby acquire sufficient energy to escape from the surface of the solid, then the process is called thermionic emission because it is brought about by heat. Thermionic emission provides the electrons which move through the vacuum in a radio valve, but not the electrons in a transistor, which always remain inside the solid material of which the transistor is made.

Another way in which electrons may be given sufficient energy to escape is by shining light onto the surface of the material. If electrons escape as a result of receiving energy from incident light, then the process is called photoelectric emission. This is the basis of many light sensitive devices using photoelectric cells.

For a material with a given work function there is a certain critical wavelength for photoelectric emission to occur. If the wavelength is too long, there will be no emission. In general, therefore, ultra-violet or blue light, which has a short wavelength, causes photoelectric emission from more materials than the longer wavelength red or infra-red radiation.

Whether an electron stays inside a solid or escapes depends on the amount of energy the electron possesses. But inside the solid the electrons have a number of permitted energy levels, and very often important properties of the material depend on which of these energy levels are occupied. Thus a material may be an insulator if its electrons are in low energy levels, but it can be made to conduct electricity if sufficient energy is given to it to raise the electrons to higher permitted levels. Furthermore, if electrons in high energy levels give up energy and fall to lower permitted levels, then the energy emitted may be useful in special ways—lasers, for instance, depend upon such emission.

We have seen that electrons are fundamental to the structure of matter. Moreover, that the part played by an electron in the structure is very much connected with the energy levels permitted to it, and with the energy it actually possesses. In particular, we have seen that the electrons flowing as electric current through a solid may give up energy to the crystal structure, so that heat and perhaps light may be given off by the material. We shall, therefore, now go on from the structure of matter to look at the nature of energy and especially at the connection between electrons, and the generation and exploitation of the more important forms of energy.

Electromagnetic Energy

Of all the forms of energy the family called electromagnetic is perhaps the most interesting, the most fundamental and the most diverse in its properties. Heat, light, X-rays, γ-rays, radio, radar and television are all part of this family. They all lie in what is called the electromagnetic spectrum, and the great difference in their properties arises from the difference in their wavelengths. Radio waves may be as long as 1500 metres, and commonly range to as low as about 10 metres. Television and radar waves range from a few metres to a few millimetres. Then, in decreasing wavelength, there are heat, infra-red, light and ultra-violet radiation. Finally, with wavelengths of 10^{-10} metres and less, come the penetrating X-rays and γ-rays. All these types of radiation are electromagnetic, and, as this radiation passes any point in space, electric and magnetic effects can be detected.

To establish ideas about wavelength and frequency it is convenient to leave electromagnetic waves for a moment and turn to the ordinary water wave, which is much easier

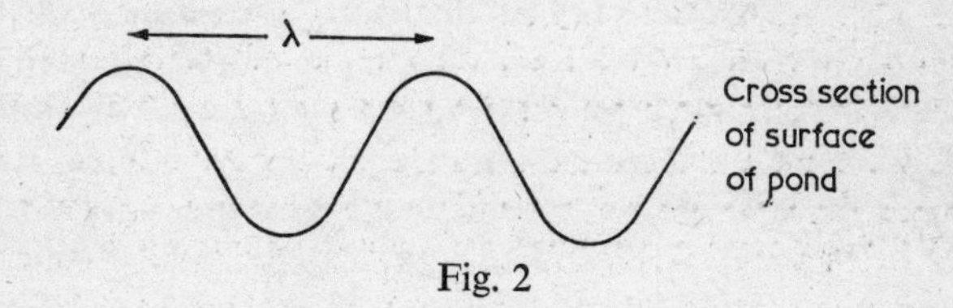

Fig. 2

to visualise because we are all familiar with the picture of crests and troughs advancing across the surface of a pond or the sea.

The number of waves passing a particular point in one second is called the frequency (f) of the wave and is sometimes expressed in cycles per second, although the term hertz is now preferred. The wavelength (λ) of the wave is the distance between two successive crests.

If f waves pass a point in one second and the length of each wave is λ, then the disturbance will have advanced a distance f times λ in one second. The velocity of the wave (c) is given by $c = f\lambda$.

If we now return to electromagnetic waves, we find that this concept $c = f\lambda$ also applies here. The velocity of all electromagnetic waves in free space is 3×10^8 metres per second, with the velocity in any material always less than this, e.g. light in glass. This particular velocity, 3×10^8 metres per second, is a very special one—fundamental to the theory of relativity, which shows that no material object, and no energy, can ever travel at a velocity greater than 3×10^8 metres per second, commonly called the velocity of light ('in free space' or 'in vacuum' often being left out of the expression).

There is a further very general concept which applies to electromagnetic radiation and which is extremely important in electronics. This is the quantum principle, which states that electromagnetic energy is only emitted or absorbed in packets of a certain size. Such a packet of energy is called a quantum, and the amount of energy (E) in a quantum is proportional to the frequency (f) of the radiation. $E = hf$, where h is a constant called Planck's constant. Several quanta may be absorbed or emitted at the same time by a substance, but a fraction of a quantum is never involved.

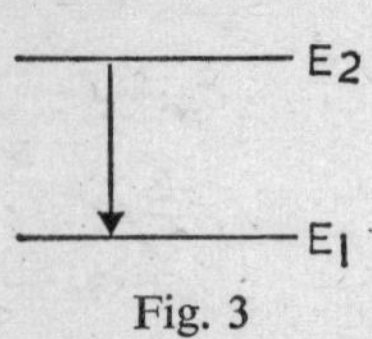

Fig. 3

The two concepts which do most to determine the way in which electromagnetic radiation is emitted and absorbed by materials are the idea of permitted energy levels for electrons and the relation $E = hf$.

In particular, if an electron loses energy it may be said to fall from a high to a lower energy level. This can be represented as in Fig. 3, where the arrow indicates that the electron which did have energy E_2 has lost some of that energy and now has a smaller amount E_1, i.e. the electron has fallen from a high to a lower energy level. If the energy that the electron loses is emitted as electromagnetic radiation, then the quantum principle is obeyed. The amount of energy emitted is $E_2 - E_1$, i.e. the size of the quantum is $E_2 - E_1$, and the frequency of the radiation is given by $E_2 - E_1 = hf$.

Atoms, molecules and solid crystals all possess their own characteristic electron energy level systems. A hydrogen atom has a certain permitted set of electron energy levels, which is the same for all other hydrogen atoms and different for all other substances.

If the atoms of an element only possess certain electron energy levels, there will be a corresponding number of frequencies for the radiation emitted by that element, because energy will only be emitted when transitions occur between this limited number of levels.

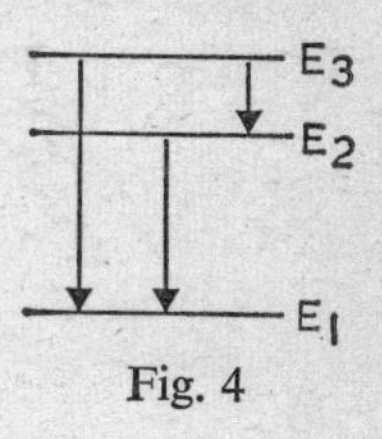

Fig. 4

Thus, if we imagine an over-simplified case where there are only three permitted energy levels E_1, E_2 and E_3, as shown in Fig. 4, then such a substance would emit radiation at only three particular frequencies, given by

$$E_3 - E_2 = hf_1$$
$$E_2 - E_1 = hf_2$$
$$E_3 - E_1 = hf_3.$$

Even the simplest of atoms, hydrogen, has many more than three permitted energy levels and consequently a larger, but still limited, number of characteristic frequencies. If a sample of hydrogen is excited in some way, e.g. by passing an electric discharge through the gas, then there are a dozen or so frequencies emitted in the optical range, together with a number in the ultra-violet and infra-red regions. The eye would respond to the mixture of different frequencies, or colours, all being emitted at the same time from a hydrogen gas discharge tube, and the brain would interpret this response by thinking that the tube emitted a predominantly reddish-coloured light. Discharge tubes containing other gases would emit different-coloured light, and such tubes are commonly used in illuminated advertisements (often called 'neon' signs, although the neon tube is only one of those employed).

The spectrometer is a more discriminating optical instrument than the eye and shows each different frequency emitted by the lamp as a separate coloured line (see Fig. 5).

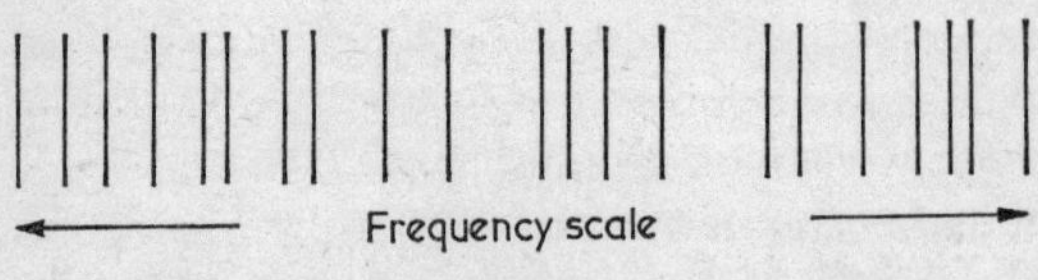

Fig. 5

A photograph of these lines is called a line spectrum, and every element when suitably excited emits its own characteristic line spectrum. Line spectroscopy—the study of such spectra—can be used as a very sensitive form of chemical detection and analysis.

Because a molecule is more complex than an atom, it has more energy levels, and thus more frequencies, in its spectrum.

If two of the energy levels in a system are very close together, then, when a transition occurs between them, the amount of energy in the emitted quantum is very small and, since $E = hf$, the frequency of the radiation is low. Instead of being up in the optical frequency spectrum, the radiation may be down in the much lower range which we call the radio-frequencies.

Certain atoms, molecules and solids can be suitably excited to emit ultra-violet, optical, infra-red or radio-frequency energy, and we shall see applications of this later. The radio-frequencies emitted in such processes are usually of the order of a few thousand megacycles per second, corresponding to wavelengths of a few centimetres.

If $f = 10\,000$ Mc/s, then $\lambda = 3$ cm, since $c = f\lambda$ and $c = 3 \times 10^8$ metres/second.

Since the term hertz is now used instead of cycles/second, 10 000 Mc/s should be written as 10 000 MHz, or 10 GHz (1 gigahertz = 10^9 c/s).

If we require radio energy at much lower frequencies, corresponding to wavelengths of tens, hundreds or even thousands of metres, then it is more convenient to produce it by causing an electron current to oscillate to and fro at this frequency in a piece of metal which is called an aerial. These methods of generating electromagnetic radiation, using electrical circuits rather than energy levels to fix the frequency, are the backbone of radio techniques and will be described later.

Let us for the present take it for granted that we can produce electromagnetic radiation at any wavelength we desire and consider now to what use we put the various parts of the spectrum.

The portion of the electromagnetic spectrum that is called heat lies between the infra-red on the short wave side and the millimetric radio waves on the long wave side.

Man, and the animals and plants upon which he depends for food, can only flourish in an environment where the temperature remains between fairly narrow limits. If the temperature of the natural environment goes outside these limits, man will first look for shelter to stop his body heat being swept away by cold winds or to shield him from the sun's radiation. When the shelter is not enough protection, the immediate environment is improved by generating heat, or in advanced societies by using heat or fuel to drive a refrigerator and reduce the temperature.

Heat was probably the first form of electromagnetic radiation generated artificially by man to improve his simple comfort and this still remains its prime function.

Light occupies a comparatively narrow portion of the electromagnetic spectrum, with a wavelength roughly between 4000 and 8000 A.U. (A.U. is short for an angstrom unit which is 10^{-10} of a metre). It is important in certain special ways—for example, in plant growth processes—but above all it enables us to see.

Information

It is important to realise the way in which we see something. If the scene is illuminated by the Sun or a lamp, then any light which enters the eye direct from the light source tells us nothing at all about the scene. It simply informs us that there is a source of light present.

The only light which carries any information about the scene, i.e. which enables us to see it, is that which has gone from the source to the objects in the scene—say, a jug on a table—and thence to the eye. The original light from the source is reflected, scattered, absorbed and generally changed by the objects before it enters the eye. This modification of the light constitutes a sort of code, which the eye and the brain decipher as a jug on a table, and when such modification of a beam of energy occurs it is said to be modulated. Selective absorption and reflection of different wavelengths give different modulation of the original light for different-coloured objects.

In the process described above, one of the sense organs receives energy, and the brain is able to recognise and extract information carried on that stream of energy. This is a one-way process, with the brain receiving information from the external world via the senses.

Suppose that, instead of acting as a receiver of information, we wanted to act as a transmitter of information. Suppose that we wanted to tell someone that there was a jug on the table. If we want to send information from us to them, then something has to be transmitted from us to them. We could speak to them and say that there was a jug on the table, which entails transmitting sound energy. We could perhaps send a postcard, and here again we note the process of modulation. A blank postcard says nothing, and it is only when we change the uniform whiteness of its surface with pen and ink that a useful message is conveyed.

If we wish to transmit the message about the jug on the table by some other method, we must first establish a link by transmitting something from us to the person with whom we wish to communicate. The first requirement is that what we choose to transmit shall safely reach the other end. The second requirement is that the transmitted energy shall be modulated to carry the detailed message we wish to convey.

Modulation

Suppose that we decide to send a message by means of a beam of light. We establish the link by turning on our lamp and shining it in the direction of the person with whom we wish to communicate. If we do nothing more, the only information he receives is that we are within his sight and have our lamp turned on. This is extremely simple information, merely saying that something is present, and as it stands the system is incapable of handling a more complex message. Under certain circumstances, however, such simple information might be exceedingly valuable—for example, if the lamp were a lighthouse standing on a dangerous rock.

Even the lighthouse beam, with its simple task of giving

warning of the presence of a dangerous rock, is sometimes modulated to carry extra information. The length and frequency of the flashes are different from different lighthouses, so that the sailor not only knows that there is a dangerous rock nearby but also, by consulting his chart, can find out which rock it is.

This type of light beam modulation is taken a stage farther in the signalling lamp commonly used for short-range communication between ships at sea. A shutter in front of the lamp is opened and closed, so that long and short flashes of light are transmitted. These correspond to the 'dashes' and 'dots' of the standard Morse code, which has a different set of dots and dashes for each letter of the alphabet. Thus any information which can be written in words may be transmitted at a reasonable speed, provided that the operators at the ends of the link are skilled.

To send messages over greater distances than about 25 kilometres, and to be unaffected by foul weather, something other than a signalling lamp is required. The range of a light beam in clear weather is limited by the horizon, but radio energy can be transmitted beyond this. If radio energy of a long wavelength is used, it will follow the Earth's curvature, and short waves may be reflected from satellites or the ionosphere, a belt of charged particles about 150 km up which bends radio signals of certain frequencies back towards the Earth. Using these methods, it is possible to establish a radio link between any two points in the world. It remains to see how the flow of radio energy between these points may be modified to carry information.

The radio energy arriving at the receiver can easily be amplified and processed to light a lamp, give a sound in a loudspeaker or make a mark on a paper tape.

Just as the light beam from the signal lamp is interrupted with a shutter, so may the flow of radio energy from the transmitter be interrupted with a key which switches the power on and off. Messages are transmitted by the operation of the key in accordance with the Morse code to produce dots and dashes at the receiver, either as flashes of light, buzzes in the loudspeaker or marks on tape.

The type of modulation which merely switches the energy on and off is rather crude, and the information which can be handled is nothing more than a written message comparatively slowly transmitted.

The transmission of much more elaborate information than this is a familiar part of daily life through the medium of radio and television. In these cases, the modulation is not simple on–off keying. It is worth considering briefly a more advanced form of modulation because certain important general principles are revealed.

On–off keying adjusts the transmitter output so that power is radiated at two levels: full power and no power. At the receiver this is converted into sound in a loudspeaker or a mark on a piece of paper when energy is being received, and nothing when no energy is received. In a typical, more elaborate, system of modulation, a constant level energy output from the transmitter produces no sound in the loudspeaker at the receiver. But if the strength of the transmitted signal fluctuates, then the receiver causes a sound to be produced by the loudspeaker.

Fig. 6 shows how the energy leaving the transmitter varies with time in an on–off keyed modulation system during the sending of a dash and dot of the Morse code. *CD* is a short pulse of energy representing the dot and *AB* is a longer pulse representing the dash, the power level being the same in each case. This is the way in which the radio energy leaves the transmitter, travels through space and arrives at the receiver.

Radio energy is part of the electromagnetic spectrum, and the dot and dash shown above can be illustrated

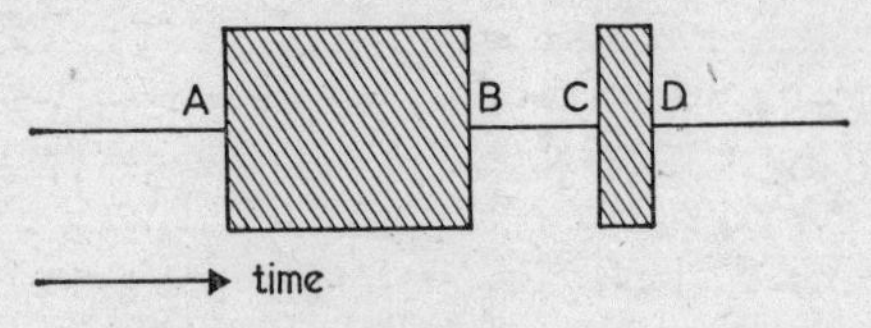

Fig. 6

again, not to show pulses of energy but the variation with time of the alternating voltage produced by the transmitter in its aerial, in space and at the receiving aerial.

In Fig. 7, *AB* again represents the dash and *CD* the dot, and the horizontal axis shows the passage of time. The wave-like curve *APQR*, etc., shows the way in which

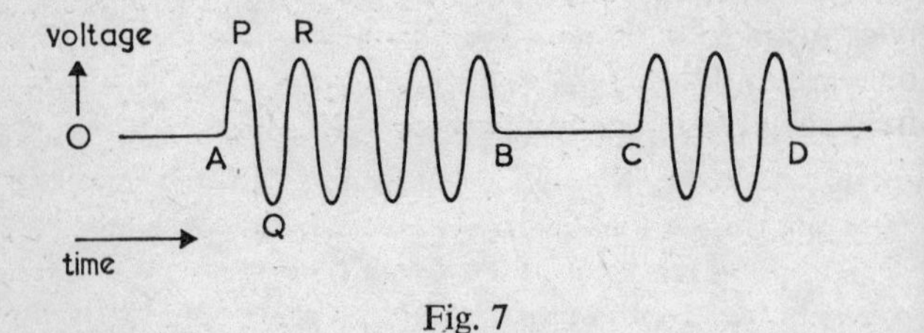

Fig. 7

the alternating voltage produced by the transmitter varies with time.

If the portion *APQR* is drawn on an expanded time scale, it can be examined in more detail.

The voltage rises smoothly from zero at *A* to a maximum value in one direction at the time indicated by *P*. It then falls smoothly to zero and increases to a maximum in the opposite direction at *Q*, before returning to zero and increasing once more to a maximum at *R*. The shape of the curve is called a sine or sinusoidal wave.

The time interval between *P* and *R* is called the

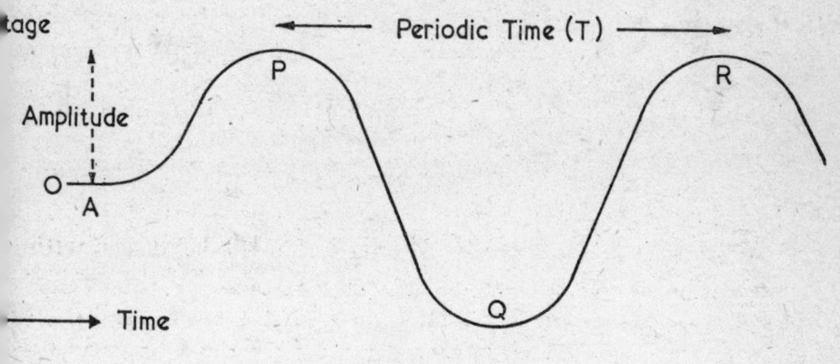

Fig. 8

periodic time (*T*), and is the time which it takes the voltage to go from one value through the whole cycle and back again to the same value. If the periodic time (*T*) is one-millionth of a second (1 microsecond), then the frequency (*f*) is one million hertz (1 MHz):

$$f = \frac{1}{T}.$$

The maximum value the voltage achieves in either direction, i.e. the value at *P* or at *Q*, is called the amplitude of the radio signal. In the case of on–off keying, the amplitude of the signal is constant whenever the key is depressed, i.e. for both dots and dashes.

Let us now look at a more elaborate modulation system

which is commonly used for the transmission of speech and music. In this particular system, no sound comes from the loudspeaker if the received signal is of constant amplitude. If, however, the amplitude varies, then the loudspeaker generates a note of the same frequency as the fluctuations in the amplitude of the radio signal. Such a system is called amplitude modulation.

When an amplitude-modulated radio signal is trans-

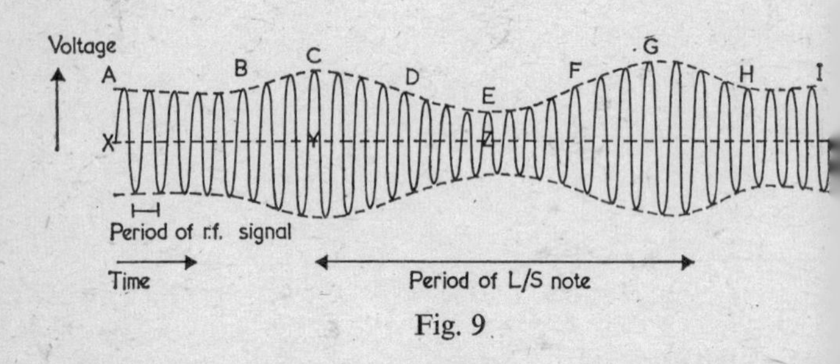

Fig. 9

mitted, the voltage may vary with time in the manner illustrated in Fig. 9. The radio-frequency voltage is shown in full line; the dotted lines are only construction lines.

Between *A* and *B*, the signal is of constant amplitude (*AX*) and the receiver would produce no sound in the loudspeaker. But between *B* and *C* the amplitude of the signal rises smoothly to a maximum value *CY*. It then falls to a minimum value *EZ* and rises again to a maximum at *G*.

During the time interval *BH*, the receiver will cause the loudspeaker to produce sound because the amplitude of the radio signal is fluctuating. A special case has been chosen for illustration in the figure where the *fluctuation* in amplitude is sinusoidal, i.e. the dotted line which just touches the crests of the signal voltage curve is a sine wave. This dotted line is called the envelope of the amplitude-modulated radio signal.

It is the function of the receiver to process the modulated signal in such a way that an alternating current of the same shape as the envelope is produced and passed through the loudspeaker.

The current corresponding to the modulation of Fig. 9 is shown in Fig. 10. The period of the note delivered by

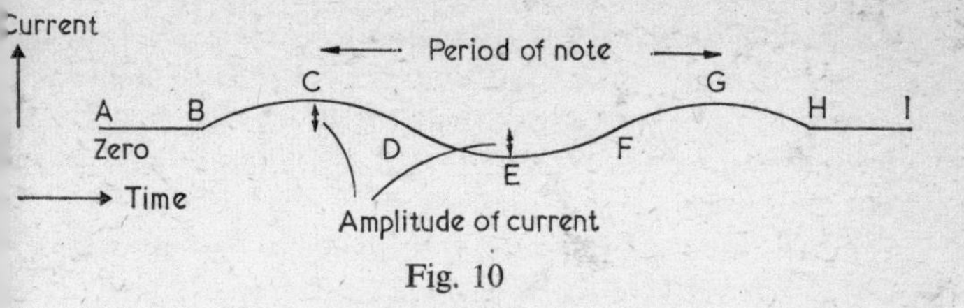

Fig. 10

the loudspeaker is determined by the periodic time of the current, and the loudness of the note depends on the amplitude of the current.

This is the type of modulation which is used a tremendous amount in radio telephone systems for short-range communication, such as between airfield control towers and aircraft, and for world-wide transmission of personal and business conversations. It is also the basis of most of the radio entertainment industry, where high-quality speech and music are transmitted.

Briefly, the sound is converted into alternating current by a microphone, and this current is used to put a modulation envelope on a radio signal. The radio signal carries this envelope to the receiver, where the alternating current corresponding to the sound is recovered and passed through a loudspeaker to regenerate the original speech or music.

This is only an outline of the theory of amplitude modulation and an account of the methods by which modulation can be carried out is given later, with some mention of other modulation systems.

If we look again at the transmission of an audio-frequency note by means of amplitude modulation of a radio-frequency wave, we can discover some important points about communication which are general principles of all systems and not just technical tricks in the circuits inside the 'black boxes'. In particular, let us consider the transmission of an audible note of frequency 1 kHz by means of the amplitude modulation of a radio signal of frequency 1 MHz.

The energy is radiated from a transmitting aerial in which an alternating current is made to flow at a frequency of 1 MHz. As long as the amplitude of this radio signal remains constant, no sound is emitted by the loudspeaker at the receiver. When the amplitude of the transmitted radio signal is made to fluctuate up and down at a thousand times per second, a 1 kHz note is emitted by the loudspeaker.

A fundamental question is why the 1 kHz note, after being converted into 1 kHz alternating current in a microphone, is not simply fed into the aerial and energy radiated directly to a receiving aerial, where the resultant 1 kHz current could be amplified and fed straight into a loudspeaker. Although such a system is theoretically possible, electromagnetic energy radiated at a low frequency like 1 kHz has a much shorter range for a given transmitter power than energy radiated at high frequencies, and such 'direct' transmission of audio-frequencies is not used in practice. Instead, high-frequency radiation which will travel the required distance is made to 'carry' the audio-frequency signal in the form of modulation.

It is fluctuation in the amplitude of the radio-frequency radiation which produces sound in the loudspeaker. Unfortunately, there are other fluctuations in amplitude as well as those deliberately imposed as modulation at the transmitter. Some of these fluctuations may be due to irregularities in the nature of the path between transmitter and receiver, particularly if it is by way of the ionosphere, which is often turbulent. But, whatever their cause, these unwanted fluctuations produce noise in the loudspeaker which tends to drown the desired signal. The signal to noise ratio (S/N) is a very important quantity in communication and anything which makes it smaller, either by reducing signal or by increasing noise, lowers the quality of the received sound. If the signal to noise ratio becomes sufficiently low, even the simplest message may be unintelligible.

If an unmodulated radio signal at, say, 1 MHz is transmitted, then an instrument testing the frequency of the radiation as it leaves the transmitter, or passes through space, or reaches the receiver will indicate—not surprisingly—that the frequency is 1 MHz. But when the signal is modulated with the 1 kHz note, the frequency meter will indicate that there are three frequencies present in the radiation: 1 MHz, 1·001 MHz and 0·999 MHz. The modulation, or change in shape, of the original sine wave gives rise to two extra frequencies, equal to 1 MHz plus 1 kHz and 1 MHz minus 1 kHz.

These three waves are normally known as the carrier wave, and the upper and lower sidebands. The carrier wave is at the original unmodulated radio-frequency, and the upper and lower sidebands are above and below this in frequency by an amount equal to the frequency of the audio note which is being transmitted as modulation on the carrier wave.

There is little point in transmitting a single 1 kHz note, but the same principle may be used to transmit a pure note of any frequency up to the audible limit and, more particularly, to transmit the mixture of pure notes which makes up most ordinary sounds. The ear will respond to frequencies up to over 10 kHz, but, if no frequencies higher than about 5 kHz reach the ear, the quality and intelligibility of sound and music is still very high.

Many domestic radio systems transmit good-quality speech and music by modulating a carrier wave with single frequencies and mixtures of single frequencies up to a maximum of about 5 kHz. This means that the maximum frequency ever present is the carrier frequency plus 5 kHz, and the minimum frequency is the carrier frequency minus 5 kHz. The total bandwidth of the system is said to be 10 kHz, i.e. from $f_c - 5$ kHz to $f_c + 5$ kHz, where f_c is the carrier frequency.

Bandwidth

The bandwidth of a system is very important in several ways. All the frequencies within the band must be dealt with in exactly the same way by the various parts of the link: transmitter, receiver and the medium in between.

If, for instance, one frequency within the band is amplified more than another by the receiver, the sound produced in the loudspeaker will not be a faithful reproduction of that transmitted. Distortion is said to occur.

When the signal is transmitted through space, there is usually little problem with such distortion, but this may arise when signals are sent along cables. It is, however, important that no other signal with a frequency within the bandwidth exists in the space by the receiving aerial.

If such a signal does exist, it will be treated exactly as though it had come from the transmitter and will represent interference or noise, tending to obscure the wanted signal.

All the elements in a system (transmitter, receiver, and cable if any) must have at least the required bandwidth. Any part of the equipment with a smaller bandwidth would limit that of the whole system to its own small value.

Generally speaking, it is more difficult, and therefore more expensive, to make broad-band than narrow-band equipment. Thus it is not surprising to find that a greater and greater bandwidth is needed as the information to be handled becomes more complicated. The cost of the equipment rises correspondingly.

This is perhaps the most important point about bandwidth—that it determines the information handling capacity of the system.

Thus a simple Morse code system might only need a bandwidth of a few hundred hertz. A system for intelligible, but fairly low-quality, speech might use just over 1 kHz, while high-quality speech and music could need, say, 9 kHz. A television link, however, will require a bandwidth of several MHz, and this must be increased still further—in principle, multiplied by three—if coloured pictures are to be handled.

Dots and dashes or low-grade speech can be handled with a fairly simple system of relatively narrow bandwidth, while television needs much more sophisticated wide-band equipment.

There is, however, a possibility of transmitting pictures over a low-grade link, provided it is done very slowly. Thus a news photograph may often be sent over a narrow band telegraph link, although it will take many minutes to transmit. Television demands the transmission of many pictures per second and thus must always have a wide band system.

The bandwidth needed in any system is determined by the complexity, or information content, of the signal and the speed with which it is to be transmitted. Conversely, a system with a particular bandwidth can handle simple information at high speed, or complex information more slowly.

If the reader thinks of himself as the receiver and a book or a teacher as the transmitter, then he will see that the transmission of information in this personal system is governed by similar laws about information content and speed of signalling. Difficult material can only be handled slowly, but simple statements can be accepted at quite a high rate.

Information Handling

We have seen that modulation of some sort is necessary if information is to be transmitted and received. The existence of so many types of modulation process—natural and invented by man—shows that the gathering and exchange of information is a very important part of human activity. It is worth considering why we are so anxious to gather information.

We get most of the information about the world around us by using our eyes and ears, and—to a lesser extent than some other animals—our nose. Less obviously, we get information about the temperature of our surroundings from all parts of the body, and about movement, especially that which affects our balance, through all sorts of things, from the soles of our feet to the fluid in the inner ear.

First of all, we need this information for our survival,

safety, well-being and comfort. As society becomes more civilized, so the need for information about the surroundings becomes less vital in terms of survival, although in time of war, when men revert to the hunters, and the hunted, the need again becomes acute, and in modern times all kinds of devices, mainly electronic such as radar, are used to extend the natural senses and help gather military information.

More highly developed societies—when at peace—are able to concentrate on well-being and comfort rather than simply on survival. In particular, amusement and entertainment become almost necessities in a sophisticated community, and this means the establishment of communication between the entertainer and the entertained. Information is transmitted from the one to the other. This is 'information' or 'intelligence', in the scientific sense rather than the everyday use of the term because the 'information' may be a joke, a song, a conjuring trick or a move on a football field.

The demand for the transmission of 'information' for entertainment has been responsible for tremendous advances in electronic techniques during the last fifty years as first radio and then television grew up. Over this period the equipment used—the 'hardware'—has become easier to operate, more reliable, less bulky and much more versatile. Fifty years ago it seemed miraculous to hear anything intelligible in the headphones of those temperamental early receivers. Now we expect the flick of a switch to give us sound of the highest quality accompanied by a good moving picture, preferably coloured.

Much of this advance is due to invention and improvement in valves, transistors and the other circuit components. But, at the same time, much has been discovered about the fundamental processes involved in any communication problem. This is discussed in Chapter 10, where it will be seen that some of these principles can be applied outside the field of telecommunications, and can tell us something about human minds and the way in which they interact.

Suppose we stop thinking about the *transmission* of information and start to consider what *use* we make of information delivered to us.

The sort of entertainment information which we get from the radio or television often affects several senses and has an immediate emotional impact. The information received when we touch a hot plate or hear a door-bell ring produces an immediate response which is almost a reflex action.

But there is some information—for example, a weather forecast or a list of train times—which we receive and retain for some time while we consider it. This 'consideration' may involve comparison with other information held in the memory and the exercise of logical deduction. Ultimately, we take a decision which is based on the information received and the result of the mental deliberation which followed.

We are now considering the use of information for a purpose which lies somewhere between survival and entertainment. This in-between region covers most of the problems we are required to solve in our working life.

A commuter may want to know at what time trains leave the local station. The manager of a supermarket chain may want to know how much bacon was sold yesterday in his shops so that he can arrange deliveries to replenish stocks. A farmer may want to know the weather forecast, perhaps to decide whether he must take precautions against frost or possibly merely to plan his day's work in a more general way.

In each case, information is in some way made available to a person so that he can reach a decision about action he will take in the future.

Typically in these problems we start with a new piece of information transmitted to us by another person. In a simple case, we then process this information in our head and decide upon some course of action.

Only too often we find that the problem we have to deal with is so complicated that, when we try to work it out in our heads, we lose the thread and get the wrong answer. If we decide to use pencil and paper, we find that problems take a long time. If we use a calculating machine, then the arithmetic is done quickly, but we still run into trouble because we take the wrong logical step in the problem, or we remember some piece of information incorrectly, or we cannot recall some important fact or arithmetical method.

The sort of problem which requires that a large amount of data be processed quickly in a set of complicated logical operations certainly overtaxes human skill, speed and accuracy. Yet this is just the sort of problem which constantly crops up in modern business and industry. Fortunately, it is also just the sort of problem which can be handled by the electronic computer.

For this reason, a new industrial revolution is taking place based on the computer, perhaps as important as the old industrial revolution which was based on the power-operated machine.

Computation

In essence, the electronic computer does all the simple things that the human brain can do, but its great power lies in its speed and accuracy. It will not carry out any process which the brain is unable to perform, but it will solve problems which involve handling simultaneously hundreds of the processes which the brain does singly. Furthermore, it forgets nothing it is told to remember, whether it is a fact or the way to do a special sum.

It is useful to remember how we solve problems ourselves because the facilities which we have in our heads will be duplicated in the computer.

Let us suppose that you are to organise a meal for the teams playing in two football matches at the local school. You will have many problems, but the first will be arithmetical.

If there are two association football matches, then there will be four teams, and each team will have eleven players, one substitute and one linesman. Food will be required for four times thirteen people—a total of fifty-two.

Here are some of the facilities you have used in this very simple start to your problem.

You have had access to a vast data store in your brain and, in particular, you have swiftly found and scanned that specialised section of the store which contains information about games, and from it extracted the figures you need about a soccer match. The information you have extracted is in numerical form and the rest of this simple problem is arithmetical.

You require to multiply thirteen by four, and the process goes something like this: four times three is twelve; write down two and carry one to the next column. Four times one is four, plus one carried from the previous column is five; write down five. Answer fifty-two.

Before this or any other sum can be attempted, a numerical system suited to the brain must be available. In our case we have a decimal system, based on the number ten. We are so used to this that we tend to overlook the possibility that any base other than ten could be used. In earlier times, other numbers were commonly used; the present British mensuration system of inches, feet and yards uses twelve and three, and many computers use two—the binary system.

In addition to the data store mentioned above, there must be a number store containing all the tables—in particular, *four* times *three*. There must also be some sort of logical facility which can test a number to see whether it is less than *ten*; if it is less than *ten* the whole number is written down, but if it is more than *ten* the last digit only is written down, the next digit being stored and added to the result of the next operation—i.e. in our sum the digit *one* from the figure *twelve* is carried to the next column. Finally, there must be some way of writing out the answer and—not so obvious with the brain, but important with a computer—some way of giving the instruction to get on with the problem.

All these facilities must be duplicated in the computer. Chapter 10, where the computer is dealt with more fully, mentions memories and stores, logic elements, and peripheral equipment for writing in instructions and printing out answers.

The idea of electronic computation was introduced in this chapter when we considered the uses made of information. Somewhere between the needs of survival and the pleasures of entertainment lie the requirements of the working world of industry and commerce, where information is used and stored and used again in solving the problems on which decisions are made about how to run offices, factories, industries and Government.

In general, all electronic computers store data and, when instructed, process it to produce answers to problems. Although all are similar in principle, some computers concentrate more on data storage than on computation and vice versa. Thus the computer designed for paying wages will differ in emphasis from one designed for mathematical research.

Apart from the pure research machines, the majority of computers are ultimately used to tell people what to do. It may be to tell the wages department how much to pay each employee; it may be to tell the operators in a continuous process chemical plant what adjustments to make to keep up the quality and the quantity of the output; it may be to tell the managing director what policy changes are most likely to produce a given effect; or it may be to tell a row of machine-tool operators what movements to make to machine a certain part.

If a computer provides a string of detailed instructions to a machine operator, then the man is only relied upon to make the adjustments, exactly as ordered by the computer.

This is a most unsatisfactory use of the man. His good points, like skill, intelligence and initiative, are not employed, whereas his bad points, such as inability to concentrate indefinitely on routine tasks, can still ruin the work. There is much to be gained by connecting the computer directly to the machine—or, more likely, to many machines—so that no human operator is needed to make the adjustments. When this is done we have what is known as 'automation'.

The control of machines by computers is only just beginning in industry, although there has been speculation for years about the effect that automation would have upon society.

The public interest in scientific discovery is so great that press and television at once give popular accounts of successful research projects. Inevitably, such articles and programmes contain speculation about the future application of any new invention. There is often a tendency for the public to get the impression that a minor scientific revolution is imminent when, in fact, a start is only just being made on the long and difficult development work which converts a promising laboratory model into a robust and profit-earning piece of factory-made machinery.

The actual construction of fully automated factories in the last twenty years hardly justifies the enormous amount of speculation that has taken place about the social and economic effects which may arise from the proliferation of such factories. Indeed, so much has been said and written about 'automation' that it has become an everyday word and is often misapplied to such things as electric tin-openers, where muscle power is simply replaced by machine power.

It is easy to accept that the control of a machine by a computer is a technically complicated process which is likely to make use of a considerable amount of electronic equipment, some of it based on novel scientific principles. It is perhaps not so obvious that the control of a machine by a human operator may also be a complex process and that sophisticated electronic equipment may be required for manual control to be performed satisfactorily.

The study of the control of machinery or processes—usually by human operators—has been going on for well over half a century. From this theory of control, or servomechanisms, have come many special electronic techniques widely used industrially, and certain new concepts—like that of feedback—which have been taken beyond engineering and are now applied to economic and social systems.

Control

Let us consider some of the principles which govern the control by a human operator of the position of a piece of machinery which is so large that a motor is required to drive it. The operator will switch on the motor when he wants the machine to move and will subsequently control the power delivered from the motor so that the machine being driven, e.g. a crane, moves in the required manner.

It will be convenient to consider a very simple system in which the operator moves a pointer and the machine is required to move to exactly the position indicated by this pointer. The operator uses a trivial amount of power to move the pointer, and the much greater power coming from the motor is controlled so that the machine being driven follows the pointer movement.

In this system, the pointer position can be regarded as the input and the machine position as the output.

One way of organising this system would be for the operator to switch on the motor to start the machine moving and then to switch off the motor when the machine reaches the new desired position.

If the motor only runs at one speed, then the operator can only switch it on when he wants to move his machine and switch it off when the new position has been reached. A two-speed motor would give the operator the ability first to move the machine quickly until it was 'nearly there' and then to switch to the lower speed for the final approach. A fully variable speed motor would allow correspondingly better control.

However flexible the arrangements for moving the machine, the system relies inherently on the operator judging when the machine has reached the desired new position.

The desired position of the machine at any time is the

input to the system, and the actual position of the machine is the output. The operator compares the output with the input and moves his machine until they are the same.

The difference between the output and the input in a control system is called the error, and it is the job of the operator to make his motor drive the machine until the error is zero.

If he has a variable speed motor, the operator will probably continually adjust the speed as the error decreases. Such a system could be made automatic if a

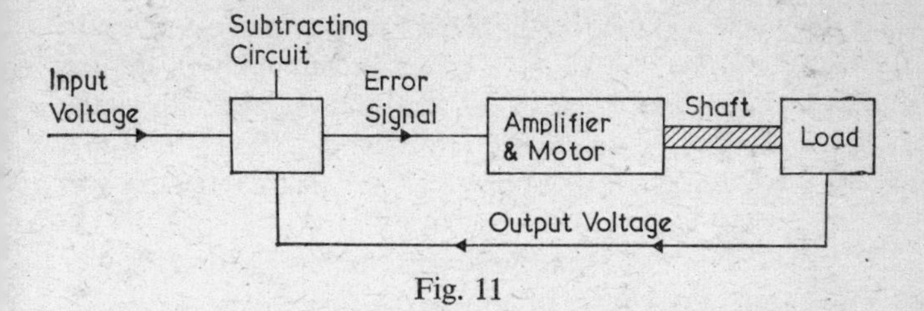

Fig. 11

method could be devised to measure the error and use it to control the motor directly, without the intervention of an operator.

When an external operator is required to observe the error and control the motor which moves the machinery, it is said to be an open-loop system. Where the error is measured and used directly to control the motor, it is a closed-loop system.

The principles of closed-loop systems are extremely important in industrial processes and also extend into biology, economics and sociology.

The elements of an electrical control system are shown in Fig. 11. The operator will move some sort of pointer and the load is required to come to the position indicated.

The pointer will be connected to a simple electrical unit like a potentiometer so that a voltage depending on pointer position is delivered from its terminals. This is used as the input to the control system.

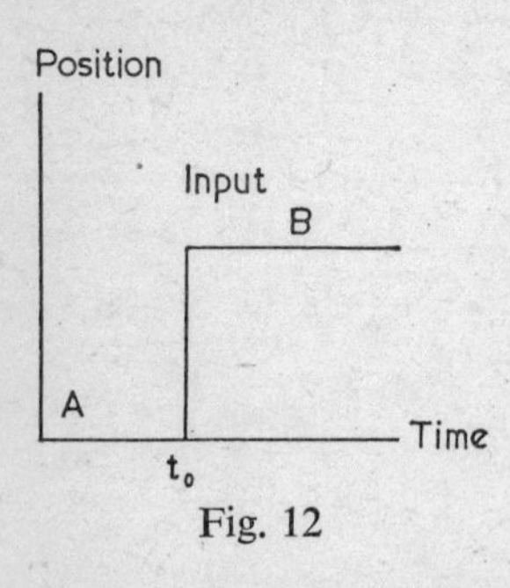

Fig. 12

A similar potentiometer is connected to the load so that a voltage is available which depends on load position. This is used as the output voltage in the system.

These two voltages are fed into a subtracting circuit, which takes one from the other and so produces the error signal. The error signal is then amplified and applied to the electric motor, which drives the load.

As long as the load position differs from the pointer position, there will be an error signal supplied to the amplifier, and the motor will drive the load to reduce the error to zero.

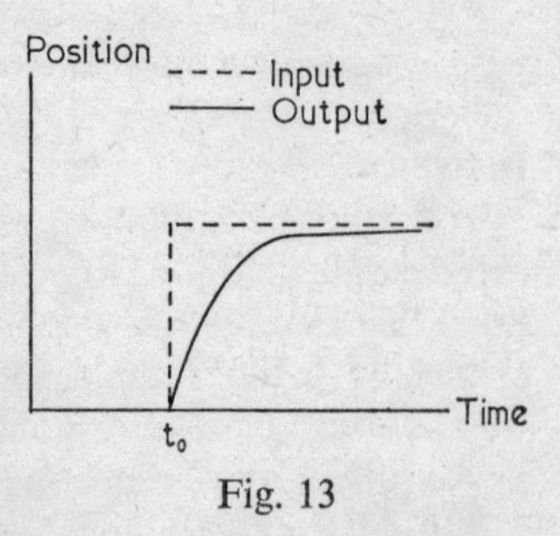

Fig. 13

Let us now examine in more detail how the load moves when the operator shifts the pointer to a new position.

The input to the system, shown in Fig. 12, is a step function, i.e. it is assumed that at time t_o the operator moves his pointer suddenly from position A to position B.

The load driven by the motor will also attempt to move suddenly from the corresponding position A to position

B. But, because of the inertia of the load and the frictional forces which arise when it moves, the system will behave typically as in Fig. 13.

In the case shown, when a step input occurs, it takes a significant time for the load to take up the new position. Towards the end of the movement the final position is being approached quite slowly.

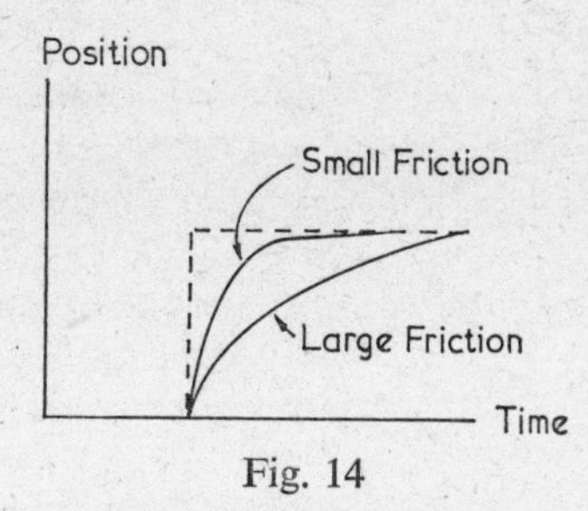

Fig. 14

Fig. 14 shows the response of a system with large, and with smaller, frictional forces. The greater the friction, the greater the delay will be before the load comes close to its new position.

To reduce power losses and unwanted heating, systems are generally designed to have as little friction as possible. So, if the response to a step input is too slow, there is little chance of further reducing friction to speed it up. Later we shall see that the response of the system can be improved in another, more subtle, way.

Although many control systems will tend to respond to a step function input in the way described above, others may have an oscillatory response.

In Fig. 15 the response to a step input is shown in two cases. In each case the load comes up to the required position and overshoots it. The error signal thus changes sign and the motor drives the load in the opposite direction, but it again overshoots the required position.

In (i) the overshoots get less each time: the oscillation eventually dies out and the desired new position is achieved. In (ii) the oscillations about the required position grow in amplitude and, unless there is some limiting safety device, such an unstable system will cause damage.

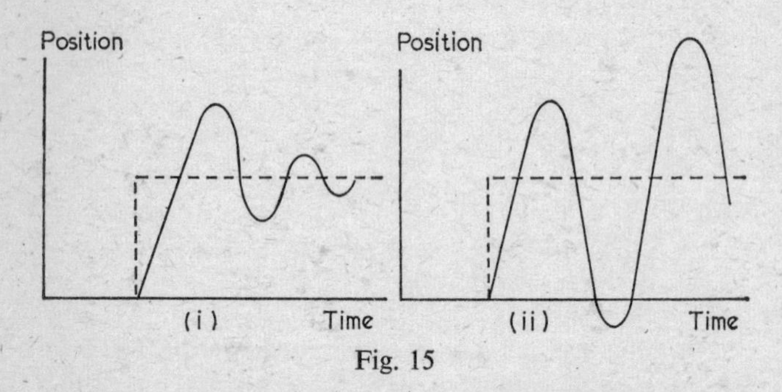

Fig. 15

If extra frictional damping is introduced into the system, the oscillatory response can be removed. But the introduction of such extra friction would lead to uneconomic loss of power and possible overheating. There is another method of correcting an oscillatory system which involves a modification to the feedback.

In the simple system so far described, an electrical signal proportional to the output is fed back to the input and used to derive the error signal, which is amplified to drive the motor. Once the load is moving, frictional forces proportional to the velocity appear and oppose the drive produced by the motor. These frictional, or damping, forces have a considerable effect upon the response of the system. In some cases the response could be improved by decreasing the friction (Fig. 14) and in others by increasing it (Fig. 15).

For the reasons stated earlier, it is not usually possible to make a significant change in the actual frictional force which opposes the driving force of the motor. Instead, an extra term is added to the feedback, and hence to the signal energising the motor. This extra term is made proportional to the output velocity, just as the real friction is.

If less damping is required in the system, then the sign

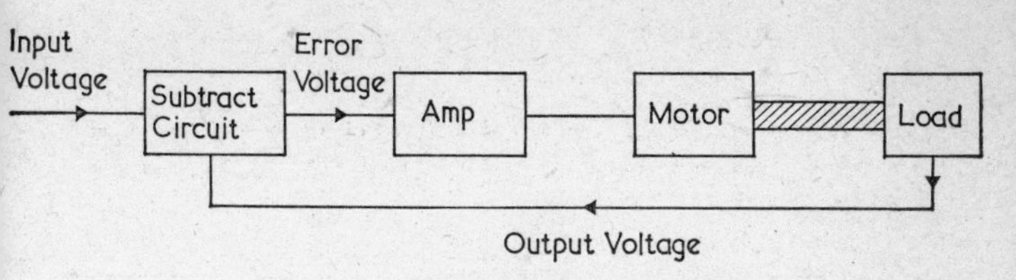

Fig. 16

of the velocity feedback term is arranged so that the signal energising the motor, and hence the drive, is increased. If more damping is required, it is arranged so that the velocity feedback term is subtracted from the signal applied to the motor. The effective driving force then falls, just as it would have done if the real frictional forces had been increased.

In the system shown above (Fig. 16) the feedback loop is simple. A device attached to the load provides a voltage dependent on the load position (devices which convert quantities like position or pressure or speed into electrical signals are called transducers). This voltage, which is proportional to the output of the system, is fed back directly and subtracted from the input voltage, which is governed by the desired position.

The signal applied to the amplifier, and hence the motor drive, is simply the error signal:

$$\text{Drive} = \text{Input} - \text{Output} = \text{Error}$$
$$D = I - O = E.$$

Such a simple system may have any of the defects of response shown in Fig. 15 and Fig. 14 when a step function input is applied. These defects could be reduced if the frictional force, which opposes the drive and is

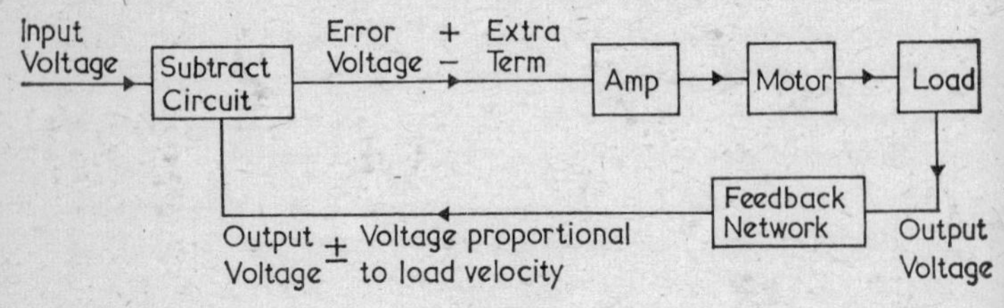

Fig. 17

proportional to the velocity of the load, could be changed.

The drive is artificially increased or decreased, as appropriate, by a term proportional to the load velocity. If a term proportional to the rate of change of output is added to the drive, this will have the same effect on the response as a change in the actual friction in the system.

In Fig. 17 the simple system has now been modified by the addition of an extra term proportional to the load velocity so that the effective friction in the system is changed to the desired value.

This is done by a change in the feedback loop so that there is an extra voltage term proportional to the rate of change of output, i.e. to load velocity.

The appropriate equation now becomes:

Drive = (Input − Output) ± Rate of change of output

$$D = (I - O) \pm \frac{d}{dt} O.$$

(Constants of proportionality have been omitted and $\frac{d}{dt}$ is mathematical shorthand for 'rate of change of'.)

The system is now said to have velocity feedback, and this is normally arranged by inserting an electrical circuit or network into the feedback path. This feedback network takes in the simple output voltage and provides a signal proportional to the output voltage plus the rate of change of output voltage.

The function of the feedback circuit is to shape the simple output feedback before it is applied to the subtracting circuit so that the desired system response is obtained.

There are other types of unwanted response and error which may arise in a control system, especially when forms of input variation with time other than a step function are used. But the process of correcting the response usually involves inserting into the feedback loop a network which will modify the simple feedback signal so that the desired response is obtained.

Sometimes a network will be inserted between the input and the subtracting circuit so that an additional term to the simple input signal is applied to the subtracting circuit. An extra component will then appear in the error signal which energises the motor, and the performance of the system will be correspondingly changed. Modification of the input in this way is sometimes referred to as feedforward.

In most engineering systems, the simple closed loop of Fig. 16 is modified by the introduction of extra circuits into the feedback or input leads in order to 'tailor' the response to some desirable pattern. In elaborate systems there will often be, in addition to the main feedback loop between input and output, sub-loops providing feedback between intermediate points.

In the previous section on servomechanisms, some of the difficulties have been discussed which arise in ensuring that the movements made by an operator are exactly reproduced by the machinery he is controlling. It is in some ways ironic that such problems also occur—and solutions have long ago been evolved—in natural biological systems.

There is still much to be discovered about the basic mechanisms of the generation and distribution of electricity throughout a living organism. Nevertheless, modern medicine utilises many empirically established ideas and techniques. These may involve putting electricity into the body, or observing the electric currents naturally present in the body.

The restoration of normal heartbeat when it has stopped or when fluttering (fibrillation) occurs, stimulation of skeletal muscles in electrophysiotherapy, diagnosis assisted by electrocardiography, and encephalography which looks at the electrical patterns in the brain—these are all common techniques using electronic apparatus which is often of some complexity.

Much research is at present directed at discovering the methods of control in biological systems, particularly the human system. There is a postural control system with position sensors, motor elements, some data processing and possibly a certain amount of adjustment of the parameters of the system by higher brain centres. There is a separate system of voluntary control of movement which is used in carrying out different tasks requiring thought and skill. In this system there is an element of adaptive control and learning.

Adaptive control means that the system slightly changes its parameters, e.g. feedback proportionality constants, during the performance of the task so that its behaviour is better suited to the particular job.

There are obvious advantages in using such adaptive principles in engineering systems, and research in biomechanics seeks to discover and adapt natural processes to industrial use. Conversely, technology can contribute to the understanding of human processes, and experiments with rudimentary machines which 'think' or 'learn' are helping to clarify theories about the working of the human brain.

Conclusion

This chapter has reviewed the wide generality of ways in which electronics interacts with daily life through technology and the mechanics of natural processes.

Both in technological and in natural systems, the initially daunting complexity of any piece of equipment or electronic process can be overcome by recognising that there are two elements involved in understanding it. First, there is the fact that the prime function can be described in ordinary language, without jargon, mathematics or circuit diagrams. Secondly, there is the fact that the most complex of systems is made up of a multiplicity of simple units—like a matchstick model of Cologne cathedral.

The purpose of the following chapters is to describe the most important of such simple units and to indicate how they can be put together into complex systems.

CHAPTER 2

Sources of Useful Electrons

Every material contains electrons, but they are not necessarily available for use in electronic devices and systems. In this sense 'useful electrons' are those which can move through materials. In doing so they produce the various effects we want, such as heat and light, or the magnetism utilised to give movement in motors and relays. But, as well as flowing through the materials of the circuit, useful electrons must be amenable to control by some external agency so that the effect they produce can be switched on and off, or otherwise adjusted, as desired.

The materials in which electrons can be made to flow are conductors and semiconductors. These will be described in this chapter. There will then be an account of the emission of electrons from solid material into a vacuum, because this is the basis of the thermionic valve. Electrons emitted from one electrode (cathode) are attracted by the positive potential on another electrode (anode). The current between them can be controlled by the potential applied to a third electrode (grid) which partly obstructs the electron path through the vacuum from cathode to anode. This is the method of operation of the valve, the well-known 'glass bottle' of radio and television sets. Important valves are described in the next two chapters, together with the solid state devices, such as transistors, in which electric current is similarly controlled, but without the electrons ever being emitted from the solid material of the circuit.

Conductors and Insulators

We have already seen (page 5) that in a single, isolated atom the electrons and nucleus are held together by electrical forces which may become quite complex if there are a large number of electrons in the atom. The constituent atoms of a molecule are similarly bound together; so, too, are the multitude of atoms in a solid crystal, which in many ways behaves like an extremely large molecule.

The cohesive forces, which bind the individual atoms into the regular crystalline lattice, characteristic of the solid state, are due to complex interactions between all the electrons and the nuclei. Some of the electrons may play a bigger part in the binding process than others, just as in a single atom some electrons are more tightly bound to the nucleus than others.

In some solids, certain of the electrons play such a negligible part in the binding process that they can be anywhere in the material and can possess any of a wide range of energies, without in any way upsetting the physical structure of the crystal.

In other substances, all the electrons are vital to the crystal structure, and neither their position in the solid nor the energy they possess may be altered.

Solids in which some of the electrons can accept extra energy and move to any part of the material are electrical conductors. The commonest examples are the metals. Substances which contain no such 'free' electrons are insulators.

When a battery is connected to a piece of metal, the electric field thus applied to the solid causes the free electrons to move towards the positive terminal. Electrons from the battery are pushed into the metal from the negative terminal to replace those entering the battery at the positive side.

The electrons drift through the conductor, drawing the necessary extra energy from the battery or any other source of electric field. The extra energy given to a free electron by the battery is comparatively small, and the drift velocity is typically about a centimetre per second.

When a switch is closed in an electric circuit, it is the electric field which is propagated through the circuit at the velocity of light. Although the electrons making up the current only move slowly, they all start to move almost at once, so the electrical effects in the circuit appear to occur almost instantaneously.

We have a fairly simple picture of the electrical nature of crystalline solids. They all have a regular crystal lattice structure, but some of them have free electrons able to move through the lattice, while others have no such free electrons.

Those with free electrons are the electrical conductors, and the energy associated with the flow of current is the kinetic energy of the free electrons due to their drift velocity. This energy is acquired from the electric field established in the material by the battery. The heating effect of an electric current is due to the free electrons giving up energy to the crystal lattice as they drift through it.

We have already seen (page 6) that electrons are often restricted in the amount of energy they may possess. If the simple picture of conductors and insulators given above is extended by including the energy level concept, then it is possible to understand semiconduction and the associated phenomena of modern solid state electronics.

Energy Level Diagrams

We have already used diagrams in which horizontal lines represent permitted energy levels for electrons. The higher up the diagram a line is drawn, the greater the energy that is represented.

In the energy level diagrams used in modern solid state theory, the same principles apply, but two sorts of horizontal line are used to denote the electron energy levels which are permitted in any particular system.

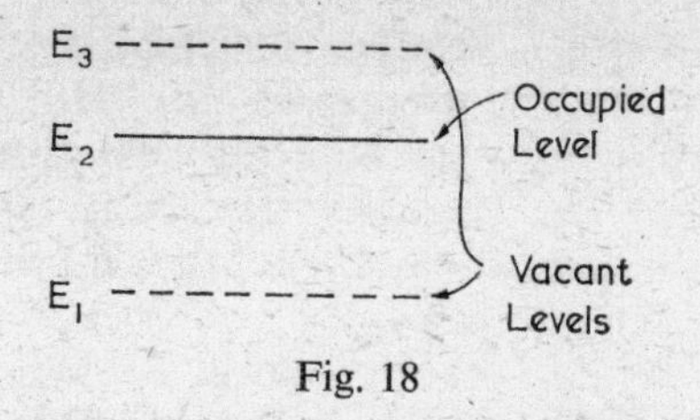

Fig. 18

In Fig. 18 the three horizontal lines represent three values of energy which an electron in some simple hypothetical system may have, $E_3 > E_2 > E_1$.

But in the system represented here there is no electron with either of the permitted energies, E_3 or E_1. These two energy levels are said to be vacant levels and are shown as dotted lines.

There is an electron in the system with energy E_2. This level is said to be occupied and is conventionally represented by a full line.

Among the rules governing electron behaviour in any system is the one which says that no two electrons can be in exactly the same state. This is called the exclusion principle, and it is incorporated into the energy level diagram convention by declaring that there can only be one electron per energy level. In principle, then, there will be a full line on the diagram for every electron in the system described, and there will also be many dotted lines.

In the system represented by Fig. 19 there are two electrons, occupying levels E_1 and E_4 respectively. There are a number of possible transitions in this material. The electron in E_4 may drop to E_3 or E_2, and if this occurs the substance will emit energy. Alternatively, the electron in E_1 may be raised to E_2 or E_3 if the substance absorbs exactly the correct amount of energy. No transition is possible between levels E_1 and E_4 because both are occupied.

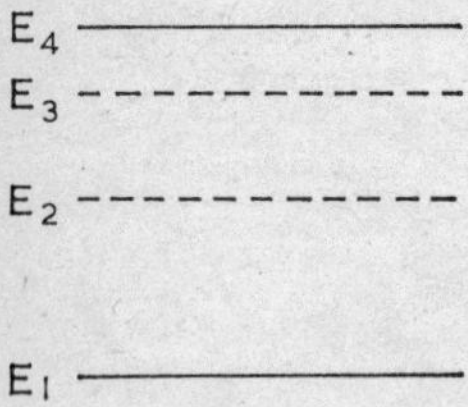

E_4 ————————

E_3 - - - - - - - - - - - -

E_2 - - - - - - - - - - - -

E_1 ————————

Fig. 19

If the energy level diagram for any crystalline solid is investigated, it will fall into one of the two patterns shown below in Fig. 20.

In (i) there is a band of many filled levels separated by

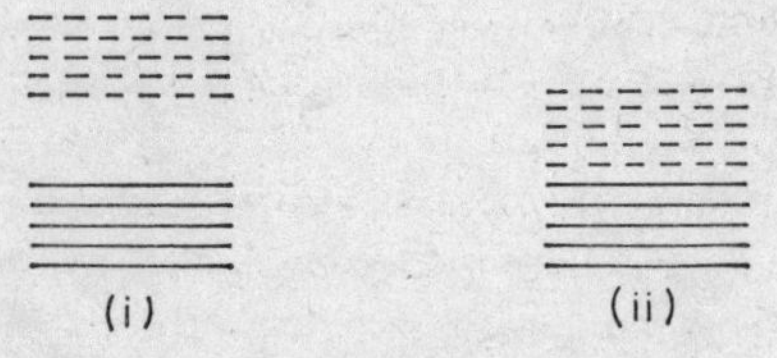

Fig. 20

a gap from a band of many vacant levels. In (ii) the completely vacant band is immediately above the completely filled band.

There will, in fact, be many more levels than this in

each band, but it is unnecessary to show them all in the diagrams used to explain solid state phenomena. Furthermore, in a complete energy level diagram there would be a number of completely filled bands well below those shown and separated by large energy gaps.

Interactions involving the electrons in such low-lying bands require very large amounts of energy and do not occur in electronic devices, except for some special ones involving X-rays or γ-rays. The amount of energy in an X-ray or γ-ray quantum may be sufficient to excite electrons out of 'deep' bands, but this is rarely relevant to ordinary electrical processes in conductors and semiconductors.

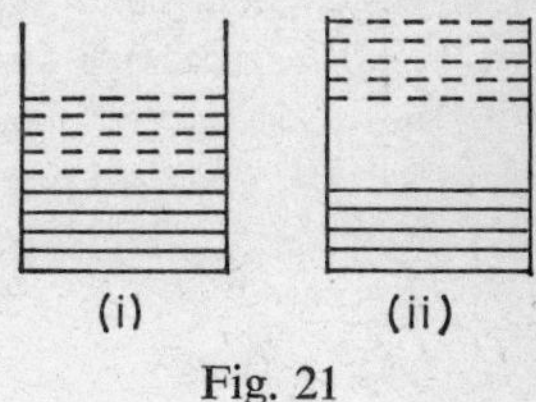

Fig. 21

The energy level diagrams that concern us are thus drawn to show only the most energetic electrons in the material. The diagrams are often drawn with vertical boundary lines at the sides; the bottom horizontal line does not, of course, represent zero electron energy.

If a material is to conduct electricity, it must contain electrons which can have their energy increased by a small amount and thus drift through the material as an electric current.

Materials with an energy level diagram like Fig. 21 (i) are conductors. There are vacant levels available down near the filled band, and the electrons are thus able to accept the small increment of energy from the battery which gives them the drift velocity.

Materials with an energy level diagram like Fig. 21 (ii) are insulators. A small increment of energy would only be sufficient to take an electron up into the gap near the filled band. But there are no permitted levels in the gap, and therefore such a small energy increment cannot be accepted by the electrons and they will be unable to flow through the material as a current.

The energy level diagrams shown above—and, indeed, all such diagrams unless otherwise stated—are drawn for materials at zero degrees absolute. At any real temperature—in particular, at room temperature, say 290 degrees on the absolute scale (290K)—the material will contain thermal energy which may affect the picture.

The effect of thermal energy upon the energy level diagram for a conductor will be to blur the sharp dividing line between the filled and vacant bands. Some of the electrons near the top of the filled band will have accepted thermal energy and will have moved up into vacant levels near the bottom of the vacant band.

Well away from the original dividing line the two bands will still be exclusively vacant or filled respectively, since the thermal energy is not sufficient to excite an electron very far up the diagram. But near this line, on both sides, there will be regions where some levels are vacant and others occupied. The material will continue to behave as an electrical conductor because nothing has altered the ability of the electrons to accept a small increment of energy from the battery.

The energy level picture of the insulator will be unchanged at room temperature because the thermal energy available would only be sufficient to take an electron up into the energy gap, and this is forbidden.

Intrinsic Semiconductors

Suppose there is a material with the characteristic insulator energy level structure, but where the forbidden energy gap is much smaller than in a true insulator. At room temperature, the thermal energy available may now be sufficient to excite a small number of electrons across the gap from the filled band into the vacant band. This is called thermal excitation.

If this occurs, the electrons in the previously vacant band can accept a small energy increment from a battery, and some of the electrons in the filled band can do the same by moving up into levels recently vacated by the thermally excited electrons which have crossed the gap.

In principle, the material has become an electrical conductor. But the number of electrons excited across the gap will be extremely small compared with the very large number which can act as charge carriers in a true conductor. The current flowing for a given applied field will be smaller than that in a true conductor by a factor of many thousands.

Materials which definitely conduct electricity, but considerably less than metals, are called semiconductors.

The material considered above, which becomes a semiconductor when raised to room temperature—germanium is an example—is called an intrinsic semiconductor.

In a metallic conductor there are always electrons present which can act as mobile charge carriers. But in the intrinsic semiconductor there are no mobile charge carriers available until thermal excitation occurs. The number of available charge carriers, and hence the conductivity, will depend much more strongly upon temperature for intrinsic semiconductors than for metals.

Impurity Semiconductors

Certain materials with the characteristic insulator structure can be converted into semiconductors by adding a small amount of impurity. Thus phosphorus added to pure silicon in the proportion of a few parts per million converts it into a semiconductor.

Such materials, which rely upon the presence of impurity to produce electrical conductivity, are called impurity semiconductors. The greater part of modern solid state electronics depends upon impurity semiconduction.

The energy level diagram for an impurity semiconductor, e.g. silicon 'doped' with phosphorus, is shown in Fig. 22.

There is insufficient impurity to upset the binding of the crystal, but the energy level diagram becomes modified, with the introduction of an extra filled level up near the vacant band, due to the impurity atom. The filled impurity level is represented in the manner shown in the figure.

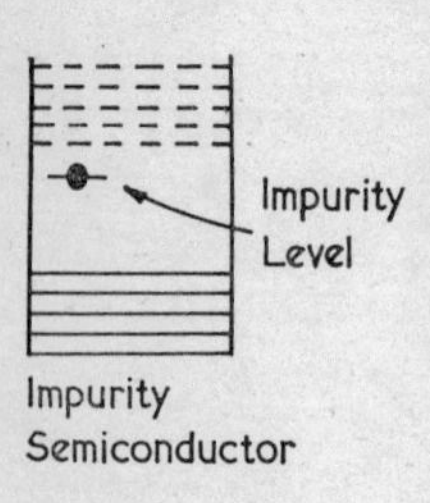

Fig. 22

The short line indicates that this is a situation which obtains only in the vicinity of the comparatively few impurity atoms, unlike the other energy levels which are permitted throughout the whole body of the crystal.

The filled impurity level is close to the vacant band and, consequently, at any temperature significantly above 0 K the electron will be excited up into the vacant band. Typically, all the electrons in the impurity levels throughout the material will have been thermally excited into the vacant band at, say, 200 K.

Once an 'impurity electron' gets into the vacant band, it may wander anywhere in the material because each energy level in the band persists throughout the whole crystal. The electron thus excited into the vacant band can accept energy from an electric field and the material is now a semiconductor. Such a material is termed an impurity semiconductor.

The conductivity will vary strongly with temperature up to the rather low temperature at which all the electrons are excited up out of the impurity level. Above that temperature the conductivity variation will be small. Under normal working conditions, the conductivity of an impurity semiconductor is largely determined by the percentage doping.

Fig. 23 gives a picture of what happens in the impurity semiconductor described above.

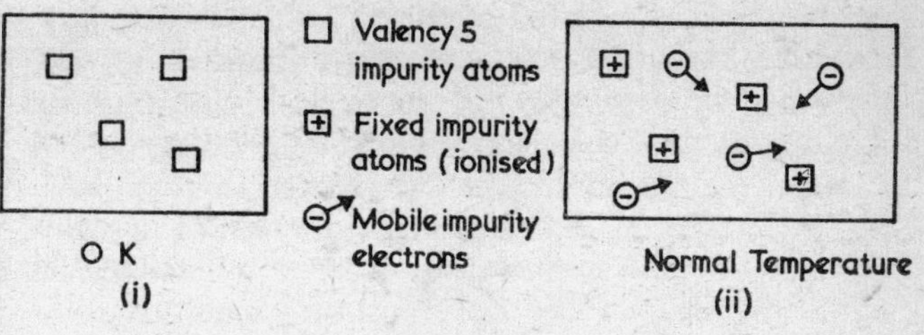

Fig. 23

The valency of the main material, silicon, is four. This means that, in pure silicon, each of the four valency electrons plays a part in binding the silicon atom into the lattice. When the impurity is added, the impurity atoms each take the place of a silicon atom in the crystal lattice. But the valency of the phosphorus impurity is five and, since only four electrons are needed to bind a phosphorus atom into a silicon crystal, there will be one electron which is not required for binding. At 0 K this electron will remain very weakly attached to its parent atom.

At normal temperatures, however, the electron will pick up thermal energy and break away from its parent atom, leaving it positively ionised. The electrons which have been thermally excited away from the impurity atoms will wander at random throughout the whole of the material. If an electric field is applied to the material, they will acquire an additional drift velocity in the appropriate direction and constitute a current.

The positive impurity ions with which these mobile electrons were originally associated remain fixed in the crystal lattice.

Another way of making an impurity semiconductor is by adding an impurity of valency three, such as boron, to the tetravalent silicon host material. There are now only three valency electrons available to bind the impurity atom into the lattice, whereas four are required. At 0 K there will be a vacancy for an electron next to each impurity atom. Such vacancies are called 'holes'.

When the temperature of the material is raised to a normal value, a nearby electron will gain sufficient thermal energy to be excited away from its parent silicon atom and occupy the 'hole'. A little later, another electron from elsewhere will move to occupy the new position taken up by the hole. This process may be thought of as the successive movement of a large number of electrons a short distance, or as the movement of the hole a long distance.

When an electron moves from left to right, it carries negative charge in that direction. To get the same electrical result when we consider the hole moving from right to left we must remember that the hole carries an effective positive charge.

The energy level diagram for a tetravalent host material doped with trivalent impurity is shown in Fig. 24 (i). There is now a vacant impurity level down near the filled band. An electron from the filled band will be excited into the impurity level thermally, leaving a vacant level in the filled band. Alternatively, it can be considered that the impurity hole has been excited *down* into the filled band.

Note that holes need to be given energy to move down an energy level diagram, whereas electrons need energy to move up.

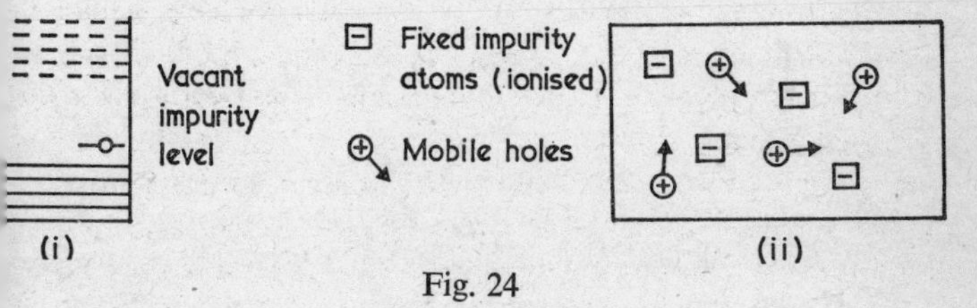

Fig. 24

In Fig. 24 (ii) is shown a physical picture of the material at a normal temperature. Holes are wandering at random throughout the whole crystal, and the original impurity atoms have now become fixed negative ions since they have each picked up an extra electron.

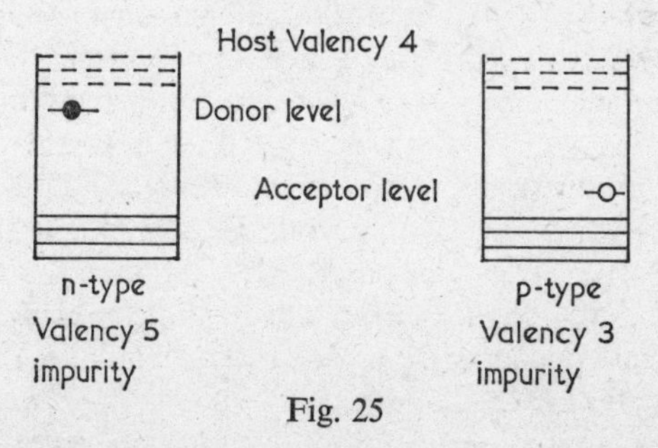

Fig. 25

Impurity semiconductors are divided into *n*-type and *p*-type, and the energy level diagrams are repeated above. The names arise from the fact that the conduction is by negatively charged electrons in the *n*-type and by positively charged holes in the *p*-type. The impurity levels in *n*-type and *p*-type are sometimes called donor and acceptor levels respectively because they give or accept an electron.

Group III–V Materials

Important semiconductors are now being made which are compounds of a trivalent and a pentavalent element, like gallium arsenide (GaAs). If the substance is prepared with an excess of the pentavalent element, it becomes an *n*-type semiconductor, while an excess of the trivalent element gives a *p*-type material.

Electronic Circuits

The simplest electric circuit consists of a battery, or some other generator of electric power, connected to a load, which may be a light bulb, a heating element, a motor or something similar. The bulk of the circuit will be a conductor or a semiconductor (usually the former) and will only be required to allow the passage of current. But somewhere in the circuit will be the point where the current is to be controlled by an external agency. At this point, there will usually be a valve or a transistor in which an electrical signal from the external agency controls the main current flowing between the battery and the load. In the valve this control depends upon the emission of electrons into a vacuum, and in the transistor upon the injection of electrons from one semiconductor into another in contact with it.

The thermionic and the semiconducting diodes depend upon the emission and injection processes respectively, and they are described in the next chapter. Such diodes are very important in modern electronics, even though, in general, they do not possess the full range of control and power amplification properties of the triode valves and transistors which are dealt with in Chapter 4.

CHAPTER 3

Diodes; Emission and Injection

The thermionic diode was invented in 1904, and this could be regarded as the date of birth of modern electronics, although by this time radio transmission over distances of thousands of miles had already been achieved with relatively unsophisticated apparatus.

The inventor of the diode valve was Fleming. He made use of the fact, first noticed by Edison, that an electric current could be made to flow across the empty space between the hot filament of an electric lamp and another metal electrode placed inside the evacuated bulb. This effect depends upon the thermionic emission of electrons from the heated metal filament.

Thermionic Emission; Work Function

The free electrons in a conductor are able to wander at random through the material. If the material is heated, these electrons will continue their random motion, but with increased energy.

If, in the course of its random motion, an electron comes up towards the surface, it will not usually escape. When it is well inside the body of the material, a free electron experiences equal forces in all directions due to the crystal lattice and the other free electrons which completely surround it. But when the electron is close to the surface, it is no longer completely surrounded and the forces on it are unbalanced.

If an electron did escape, the solid would have a positive charge, and an attractive force would exist between this positive charge and the negative charge on the electron.

The result of both these effects is that an electron just inside or just outside the surface will experience a force directed back into the material. This force must be overcome if the electron is to escape. At room temperature, only a negligible number of electrons approaching the surface would have enough energy to escape. But if the temperature is raised sufficiently, then the average energy of the electrons increases and the faster-moving ones are able to escape from the surface of the solid.

The minimum energy that an electron must possess in order to escape is given by ϕe, where e is the charge on the electron and ϕ, measured in volts, is the work function of the material.

In quantum terms, the work function is sometimes defined as the difference between the energy which represents the boundary between the filled and the vacant bands at 0 K, and the energy possessed by an electron just free from the conductor (see page 53).

The lower the work function, the greater the thermionic emission current obtained at a given temperature will be. The higher the temperature, the greater will be the emission.

The metal tungsten was originally used as a thermionic emitter. Although it has a high work function, it also has an extremely high melting point and could thus be operated at a very high temperature to give copious emission. Modern valves use oxide-coated cathodes, which give good emission at a much lower temperature than is needed with tungsten. The cathode material is a mixture of the oxides of metals such as barium and strontium. Free barium or strontium diffuses to the surface when the material is heated, and emission takes place from patches of metal only an atom or two thick and having a very low effective work function.

The electrons are emitted from the cathode which, in a modern valve, is usually of the oxide-coated, indirectly heated type shown in Fig. 26.

The cathode itself is a nickel sleeve coated with the mixture of oxides. The temperature of the sleeve and the oxide coat is raised by an electric heater, embedded in some sort of cement which gives good thermal contact but which is electrically insulating. Three electrical connections must be brought out through the valve envelope: the two heater terminals and the cathode lead itself.

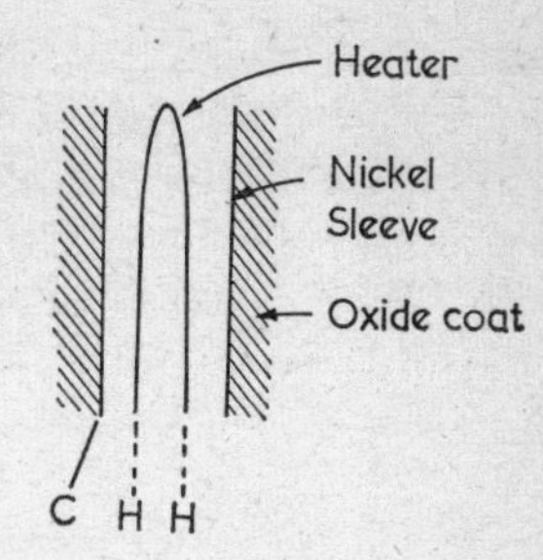

Fig. 26

Thermionic Diode

If two separate pieces of metal are connected to opposite terminals of a battery, an electric field will be produced in the space between them. An electric field is a region where a charged body experiences a force tending to make it move. If the space between the two pieces of metal is evacuated so that there are no obstructing gas molecules, an electron placed in the field will move from the negative to the positive metal electrode.

In the thermionic diode, electrons emitted from the hot cathode are attracted across the intervening space to another electrode—the anode—by an electric field produced by a battery connected so as to make the anode positive with respect to the cathode.

The electrode arrangement is shown in Fig. 27 (i), the anode being a cylindrical metal sheet surrounding the cathode and some distance from it. The electrical connections to the anode, cathode and the two ends of the heater are brought out to metal pins, which pass through the glass envelope at the base of the valve. There are insulating supports inside the envelope to carry the electrode structure. Before the envelope is sealed, the valve is evacuated to a very low pressure, and elaborate methods are used to remove gas absorbed in the metal and glass parts.

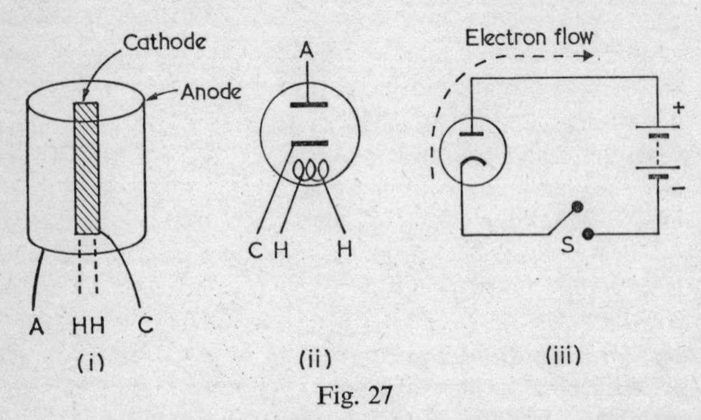

Fig. 27

The symbol used for the thermionic diode in electronic circuit diagrams is shown in Fig. 27 (ii). In such circuits there will usually be many valves and, to avoid making the diagram too complex, the wires connecting the valve heaters to the low voltage supply are not shown. In some cases the heaters are not even shown in the valve symbol, it being taken for granted that each valve has a heater and that all the heaters are connected in parallel to the heater voltage supply—usually a winding, on the main transformer in the equipment, delivering 6·3 volts a.c.

Fig. 27 (iii) is the circuit diagram for a diode valve in series with a switch and a battery. (Note that the cathode heater has not been shown.)

When the switch is closed, the electrons being emitted from the hot cathode will be swept across to the anode by the electric field produced between cathode and anode by the battery. If the switch is opened, then the field disappears and the electrons no longer cross the valve.

When the switch is closed, a current of electrons flows round the circuit in the direction shown.

But if the battery had been connected so that the anode was negative with respect to the cathode, then the electric field would be trying to push electrons across the valve from anode to cathode. There are, however, no electrons emitted from the cold anode, so no current can flow.

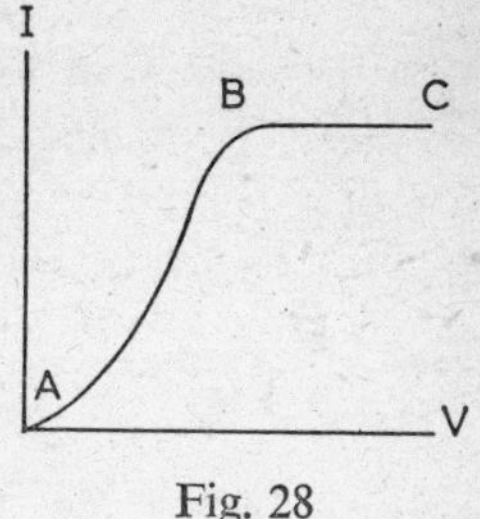

Fig. 28

The essence of diode operation is that the valve passes current when the anode is positive with respect to the cathode and passes no current if the anode is negative with respect to the cathode. The valve conducts one way only.

If a battery is applied to a diode valve in such a way that the voltage across the valve can be gradually increased, then the graph of current (I) crossing the valve, plotted against the positive voltage of the anode with respect to the cathode (V), has the form shown in Fig. 28.

If the cathode is heated and there is no positive voltage on the anode, the electrons emitted from the cathode surface will form a cloud in its vicinity called the space-charge. The negative charge on the cloud will tend to prevent more electrons escaping from the cathode surface. When the heater is switched on, electrons are emitted from the cathode until the space-charge quickly attains a value such that no further emission takes place. Under these circumstances, when V is zero, then so is I.

If V is now gradually increased, some electrons are swept across to the anode from the space-charge cloud and then immediately replaced in the cloud by others emitted from the cathode. As V gets bigger, the number of electrons per second swept out of the space-charge, and hence the current (I) in the circuit, increases. This is the region AB of the I–V curve, which is called a diode characteristic.

The cathode has a certain temperature, work function and area, and the maximum number of electrons per second that it can emit is determined by these quantities. When the current crossing the valve reaches this value, any further increase in V cannot produce an increase in I. This is the region BC of the characteristic.

The region AB is called the space-charge limited region of the characteristic, and in this region $I \propto V^{\frac{3}{2}}$ approximately. The region BC is called the temperature limited region of the characteristic because, in principle, the only way to get more current would be to increase the cathode temperature so that I was able to get to a higher value before the characteristic flattened out, or saturated. In practice, diode valves are almost invariably designed so that they are operating in the space-charge limited region.

The *p*–*n* Junction

The work function, or barrier, which prevents the escape of electrons from the surface of a conductor is the basis of the thermionic diode, where electrons can only pass from the hot to the cold electrode. If a *p*-type and an *n*-type semiconductor are put in contact, a similar barrier will be formed at the junction. This barrier tends to prevent the passage of electrons from the *n*-type to the *p*-type material and of holes in the opposite direction. Such a barrier is the basis of the *p–n* junction diode which, like the thermionic diode, conducts one way only.

We have already seen (page 56) that an *n*-type semiconductor at room temperature (say, Si doped with P) consists of a fixed crystal lattice of silicon atoms in which a very small number of the places are filled by positively charged phosphorus ions, which also remain fixed. The electrons which were attached to these ions in the original phosphorus atoms wander at random through the whole of the material.

In the *p*-type semiconductor (say, Si doped with B) there will be positive holes wandering at random throughout the silicon lattice, which will contain a few fixed negatively charged boron ions.

Let us consider what happens when a piece of *n*-type and a piece of *p*-type silicon semiconductor are in contact.

The electrons in the *n*-type semiconductor are free to wander through the material. They have been thermally excited from the localised impurity levels, which are due to the phosphorus, up into the vacant band which is a feature of the silicon host material. The *p*-type material is also mainly silicon so that, when a *p–n* junction is made, the electrons in the *n*-type will see an extension of the vacant band and will thus *tend* to wander freely in the *p*-type material—just as they do in the *n*-type. Similarly, the wandering holes in the *p*-type semiconductor will *tend* to diffuse into the *n*-type semiconductor when the two materials are placed in contact.

However, this tendency of the electrons and holes in the *n*-type and the *p*-type materials respectively to diffuse throughout the whole volume of the combined materials is opposed by two effects, described below, which arise in the vicinity of the area of contact.

A hole is a place where an electron is needed to complete the crystal binding. Thus, if a wandering hole meets a wandering electron, the electron will be captured by the hole to complete the binding, and neither hole nor electron will subsequently be available to carry current in the material. If holes and electrons meet, they destroy each other as charge carriers. This is one effect that takes place when the *n*-type and *p*-type materials are put into contact.

If an electron does pass out of the *n*-type semiconductor, it will leave the material positively charged, and when it enters the *p*-type material it will give this material a negative charge. The fact that the *n*-type material has acquired a positive charge and the *p*-type a negative charge will tend to prevent the natural diffusion of electrons from *n* to *p*. The passage of holes from *p* to *n* will cause this charge on the two sides to be further augmented.

When the two materials are put into contact, electrons and holes will diffuse across the junction until a sufficiently large charge has been built up to inhibit further diffusion. This is the equilibrium state.

The charges on the *n*-type and *p*-type materials cause a potential barrier across the junction which prevents the passage of any more of the available charge carriers across it.

If a battery is applied to the junction with its positive terminal to the *n*-type and its negative terminal to the *p*-type, the external voltage increases the height of the potential barrier still further and no current flows. If the battery is applied the other way round, the height of the potential barrier is reduced and the charge carriers are now able to get across it, i.e. current flows.

The mechanism of this 'one-way flow' across a *p–n*

junction is discussed more fully, and in energy level terms, below.

In Fig. 29 is shown the distribution of charge in the neighbourhood of a *p–n* junction. To the left of *A* and to the right of *B* is seen the charge distribution which would have existed in the *p*-type and the *n*-type materials respectively if there had been no contact. In each case,

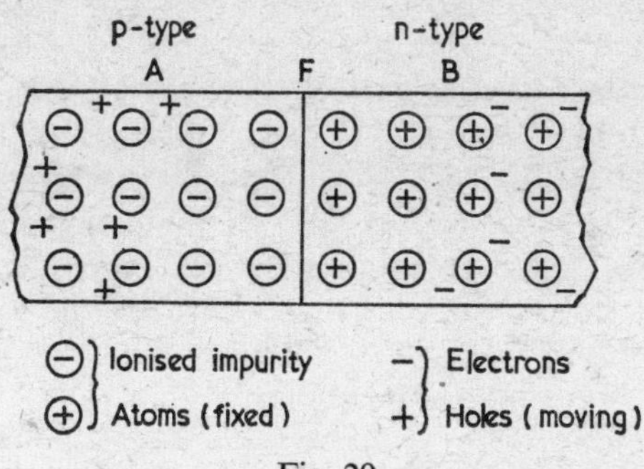

Fig. 29

the charge due to the fixed ionised impurity atoms is just neutralised by the opposite charge of the holes or electrons which move at random throughout the crystal.

In fact, however, the electrons which leave the *n*-type material leave it positively charged, and the holes which leave the *p*-type material leave it negatively charged. They diffuse into the new material until they meet a carrier of the opposite sign. When an electron and a hole meet, they combine and both are removed from the electrical conduction process. The immediate vicinity of the junction, say between *A* and *B*, contains no free holes or electrons and is an insulator across which there exists a potential gradient due to the space-charge of the ionised impurity atoms, now no longer neutralised by free carriers. The region between *A* and *B* is called the depletion layer.

Let us now look at the *p–n* junction in terms of the energy level diagram.

At room temperature, the electrons moving at random in the *n*-type material will possess a range of energies. Most of the excited impurity electrons will be in energy levels low in the originally vacant band. The higher the position in the vacant band, the fewer will be the number of energy levels filled by impurity electrons.

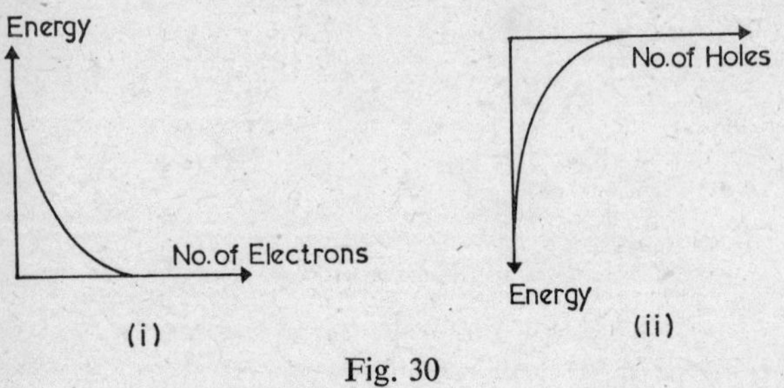

Fig. 30

A graph of the distribution of impurity electrons in the vacant band will have the exponential form shown in Fig. 30 (i). Similarly, the distribution curve for free holes in the filled band of the *p*-type material will have the form of Fig. 30 (ii), where, because we are dealing with holes, energy increases in a downward direction.

These distribution curves for electrons and holes may now be transferred to the energy level diagram to give information about the state of affairs in *n*-type or *p*-type material at room temperature, before and after contact.

In Fig. 31 the *p*-type and the *n*-type energy level diagrams are shown at room temperature immediately before the materials have been put into contact along the line XY.

The dotted line E represents the bottom of the vacant level band in the host material (e.g. silicon), and it will be common to both n and p sides. Similarly, the full line F is the top of the filled band. The shaded curves represent the impurity electron and hole distributions. The most energetic electron has the energy indicated by A and the most energetic hole has energy indicated by B.

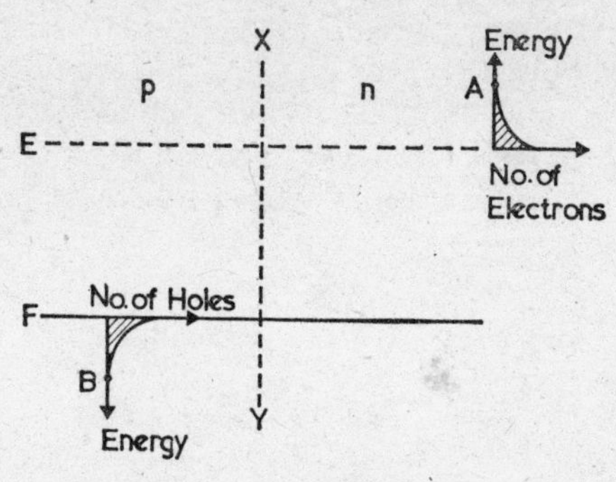

Fig. 31

Before we look at what happens when the two materials are put into contact, let us consider briefly a fundamental point about an electron energy level diagram.

Such a diagram for any particular material shows the permitted energy levels for electrons in the material and which of these levels are actually occupied by electrons. In fact, there are more energy bands than appear in the diagrams we use to describe conductors and semi-conductors, but these are all filled bands, lying well below the ones shown and playing no part in the electrical processes that interest us (see page 53).

In these processes, we are concerned with the presence or absence of electrons in the bands shown and possibly with transitions between these bands. The absolute numerical value of the energy associated with any particular level is of little interest. Much more important is the difference in energy between any two particular levels.

The energy denoted by the levels in the diagram is the sum of potential and kinetic energy, and is determined largely by the physical structure of the material. If a positive or negative electrical potential is in some way applied to a piece of material, the structure will not change, nor will the relative positions of the energy levels in the diagram. But the absolute value of all the permitted levels will be lowered or raised, depending on the sign of the applied potential. Since we are not concerned with absolute values, this will not affect any discussion of the behaviour of the material on its own.

The effect becomes important, however, if we consider two materials in contact with a difference of potential established between them. The energy levels on one side of the junction will all go up a certain amount, while those on the other side will all go down the same amount.

In the *p–n* junction described below, this effect gives a potential hill at the junction and in this way the unidirectional conduction, or rectifying action, of the junction diode arises.

Fig. 32 (i) is a repetition in simple form of Fig. 31, and shows the distribution of electrons and holes at room temperature just before the *p*-type and *n*-type materials are put into contact.

Fig. 32 (ii) shows the situation after contact, when the events described earlier (see page 68) have taken place. The shaded region in the diagram indicates the depletion layer, and the fixed charges distributed throughout tha layer will make the p-type material on the left negative with respect to the n-type material on the right. The energy levels on the left will move bodily upwards and those on the right downwards to give the potential hill between the two materials as shown.

The direction and amount of movement of the energy

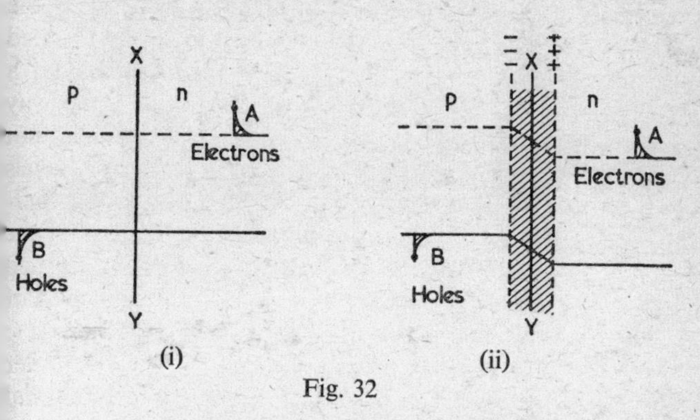

Fig. 32

levels is such that the most energetic electron A and the most energetic hole B have just insufficient energy to surmount the hill and pass across the junction.

When the two materials are put into contact, diffusion of holes and electrons across the junction takes place until the equilibrium situation described above is attained.

Let us now consider a battery applying an external potential difference between the two materials, as shown in Fig. 33.

In (i) the battery accentuates the effect due to the internal potential differences across the junction. The energy levels move and increase the height of the hill. The most energetic electrons and holes now have far too little energy to surmount the bigger hill. None of the electrons in the n-type material, nor the holes in the p-type material, has sufficient energy to cross the junction, so no current flows.

In (ii) the externally applied voltage reduces the height of the hill so that the electrons and holes ringed on the distribution curves now have sufficient energy to cross the junction. Current will flow in the circuit.

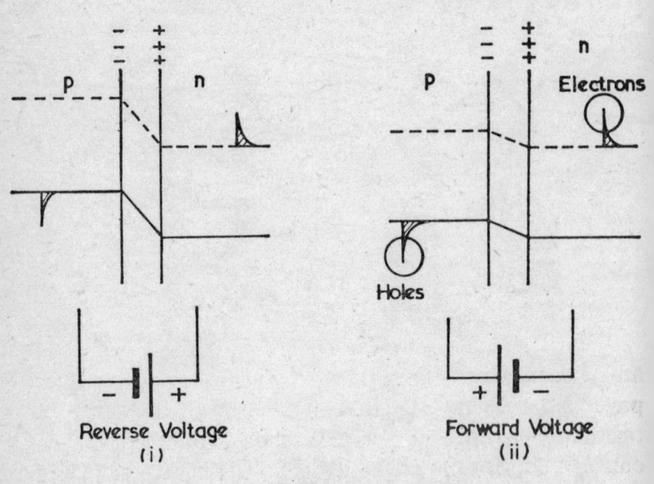

Fig. 33

As the applied voltage from the battery is increased in this forward direction, the amount of the electron (or hole) distribution curve poking over the hill will increase, and so will the current in the circuit because there are more charge carriers available to cross the junction. Since the distribution curves are exponential in shape, the current in the circuit will increase exponentially with the applied voltage in this forward direction (see Fig. 34 (i)).

In certain cases, the *p–n* junction characteristic may look like Fig. 34 (ii). The forward characteristic is not significantly different, but when a voltage is applied in the reverse direction a small reverse current flows. This effect arises when the host material used to carry the *n*

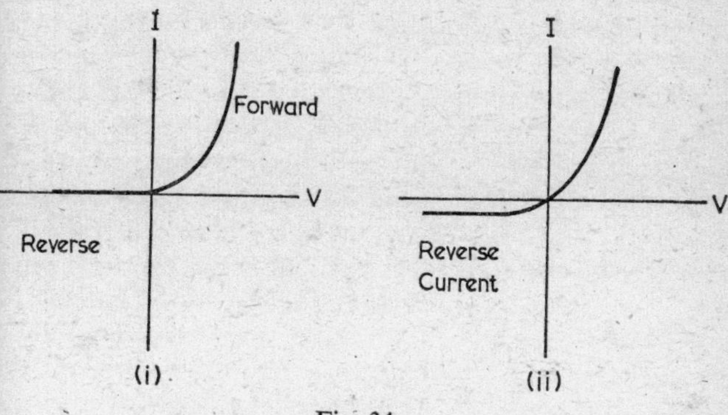

Fig. 34

and *p* impurities is itself an intrinsic semiconductor (see page 54). Thus germanium rectifiers, especially if operated at a high temperature, show significant reverse currents because the forbidden gap between the filled and the vacant bands is comparatively small. In particular, at about 70°C the effect of intrinsic semiconduction is sufficient to hide the impurity semiconduction effects which give rise to the rectifying action of the junction.

Thus germanium diodes and transistors would not be used at much above room temperature. But, because the forbidden energy gap in silicon is greater than that in germanium, the amount of unwanted intrinsic semiconduction is much less at any given temperature; in particular, it is insignificant at room temperature.

p–n Diodes

In describing the action of the *p–n* junction, it was implied that two separate pieces of material, one *n*-type and one *p*-type, were brought into contact. In fact, *p–n* junctions are normally made by some sort of diffusion, alloying or deposition technique. Impurity may be added, say, to a slice of *n*-type material in order to convert it to *p*-type. The process is stopped before it is complete, so that one face of the material is *n*-type and the other *p*-type. Electrical connections are then made to these two faces and the *p–n* junction within the body of the material gives the required rectifying action.

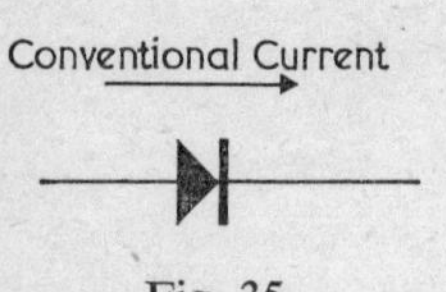

Fig. 35

The slice of material used may be only about a millimetre square if it is not required to handle much current. Thus the diodes used in their thousands for switching and logic in computer circuits (see page 165) can be extremely small, while those used for high power rectification—e.g. in electric traction systems—may be as big as a man's fist.

The symbol for a solid state diode is shown above (Fig. 35), with the arrow indicating the direction in which conventional current flows. (A conventional current flow is assumed to be made up of positive charge, but is really a flow of negatively charged electrons in the opposite direction.)

Commercial Diodes

Junction rather than thermionic diodes are tending to be used in most applications. They are smaller, cheaper and more robust, and usually operate with greater efficiency, i.e. the power dissipated uselessly as heat is less in a solid state than a thermionic diode. On the other hand, thermionic diodes can withstand much higher voltages applied in the reverse direction without breaking down and can be operated at fairly high temperatures.

The applications of the rectifying action of the diode are discussed in Chapter 9. Certain solid state rectifiers utilise a special contact between a metal and a semiconductor, which gives rise to a junction with properties like those of the *p–n* diode. Other diodes can be adapted for special applications unconnected with the rectification property, and these are mentioned briefly in Chapter 11.

CHAPTER 4

Triodes

Diodes have two active electrodes: the cathode and anode, separated by a vacuum in the thermionic valve; and the *p* and *n* material, separated by the depletion layer in the junction diode. Each diode has two wire leads, or terminals, which are connected into the circuit (the two heater leads in a thermionic valve are not counted as part of the electronic circuit proper).

Triodes have three electrodes and three wires or terminals connecting them into the circuit. In general, two terminals (say, 1 and 2) are connected into one circuit and two terminals (say, 1 and 3) are connected into another circuit, i.e. terminal 1 is common to both circuits. Electrical effects in one circuit are then used to control the flow of current in the other and, in particular, amplification may be achieved.

Thermionic Triodes

In the triode valve, the third electrode is called the control grid—often just the grid—and it is a wire spiral wound close to the cathode, but not touching it (see Fig. 36 (i)). Electrons leaving the cathode and crossing the valve to the anode must pass through this grid spiral. Fig. 36 (ii) is a half section of the electrode structure, showing that the grid is very much closer to the cathode than is the anode. Because of this, the electrical potential of the grid has much greater control than the anode potential on the electron current leaving the cathode to cross the valve.

In Fig. 37 a triode is connected so that a potential difference V_a can be maintained between anode and cathode by the high tension (h.t.) battery P, and a

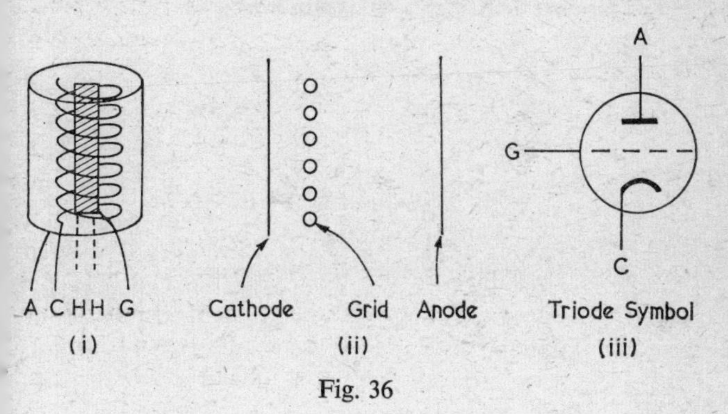

Fig. 36

potential difference V_g can be maintained between the grid and the cathode by the battery Q.

If the cathode is connected to earth as shown, then its

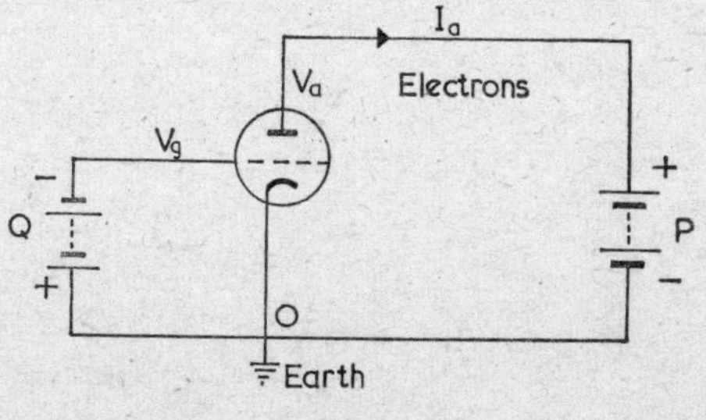

Fig. 37

actual potential is said to be zero. The control grid and anode potential are then V_g and V_a. The batteries are normally arranged so that V_a is positive and V_g negative, as shown. The current crossing the valve is I_a, and in the diagram the arrow indicates the direction of electron flow.

Let us examine a set of experimental results which might be obtained with this circuit.

V_g	V_a	I_a
volts	volts	mA
0	0	0
0	100	20
−2	100	0

With all the electrodes earthed, i.e. at zero potential, no current flows. If the anode is now put up to 100 volts, with the grid maintained at 0 volts, the current flowing is 20 mA. If the grid is now put down to −2 volts, with the anode maintained at 100 volts, the current again becomes zero.

Starting with $V_g = 0$ volts, $V_a = 100$ volts and $I_a = 20$ mA, it is possible to reduce the current to zero in either of two ways. The anode voltage can be changed by 100 volts to zero, keeping the grid voltage constant; or the grid voltage can be changed by 2 volts to −2 volts, keeping the anode voltage constant.

A given change in I_a can be produced either by a change in V_a, $\Delta V_a = 100$, or by a change in V_g, $\Delta V_g = 2$:

$$\frac{\Delta V_a}{\Delta V_g} = \frac{100}{2} = 50.$$

Thus the grid is fifty times as effective as the anode in controlling the valve current. This factor (50) is often called the amplification factor of the valve (μ), and its value is determined by the valve geometry.

A change in V_g causes a change in I_a (see Fig. 38). If a load of resistance R is placed in series with the valve and the battery, any change in the current flowing across the valve will produce the same change in current in R and thus alter the potential difference across the load R.

The end B of the resistor R is connected directly to the positive terminal of the battery. B is thus fixed in potential. If the potential *difference* between B and A changes by a certain amount, the actual potential at A will change by the same amount.

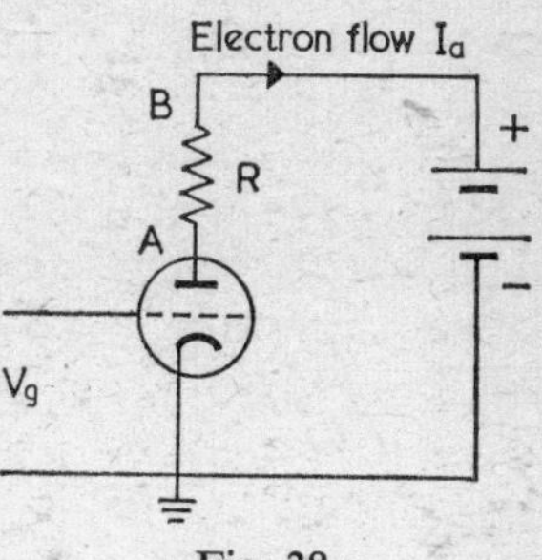

Fig. 38

The circuit shown in Fig. 38 is a rudimentary amplifier.

If the signal to be amplified is arranged to cause a change in V_g, the current I_a, flowing in the valve and in the resistor R, will also change. In particular, the potential at A will change and, if the amplifier is well designed, the change in voltage at A will be many times the change in V_g. The signal voltage has been amplified, and the maximum amount of amplification that can be obtained is theoretically equal to μ, although this cannot be achieved in practice. This is because the presence of R in the circuit reduces the change in I_a for a given change in V_g to a value below that which would have been obtained with the valve alone.

When designing an amplifier, it is necessary with any particular valve to choose a suitable value for R and a suitable voltage for the battery. These matters are discussed more fully on page 104.

In the triode amplifier, the comparatively small power

associated with the input signal is applied in the grid circuit so that it controls the much larger amount of power flowing in the anode circuit and, in particular, the power flowing in the load *R*.

Tetrode and Pentode Valves

Certain features of the triode valve made it unsatisfactory for amplifying signals which vary at very high frequency, i.e. in the MHz range. These disadvantages have been overcome by inserting extra spiral grid wires between the control grid and the anode.

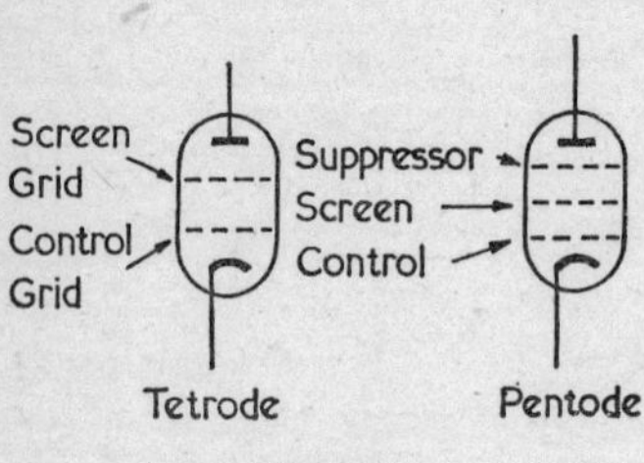

Fig. 39

First, the screen grid was added and the tetrode valve produced. This overcame the difficulties of very high frequency amplification, but the current in the tetrode had certain anomalies which led to the introduction of the fifth electrode—the suppressor grid—and the invention of the pentode.

Apart from being satisfactory up to very high frequencies, the pentode has other advantages over the triode associated with the shape of the current–voltage curves—the 'characteristics' discussed in the next chapter.

For this reason, the pentode is much more commonly met in valve circuits than the triode, although its principle of operation is the same—viz. the control of the current in the anode circuit by the application of a signal to the control grid.

The Junction Transistor

The junction transistor consists of two *p–n* diodes in such very close proximity that the current flowing across one junction affects the current flowing across the other. The two junctions are actually formed in the same piece of semiconducting material, and the charge carriers crossing

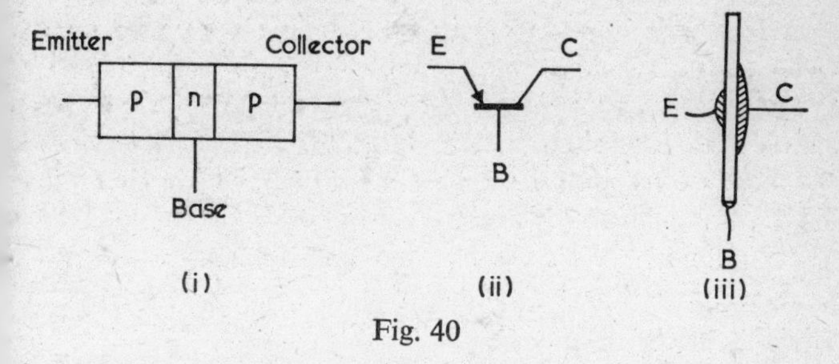

Fig. 40

one of the junctions are able to diffuse through the intervening solid material to the other junction.

Fig. 40 (i) shows the arrangement of semiconducting material in the *p–n–p* junction transistor, and Fig. 40 (ii) is the symbol used in electronic circuits. A single piece of material is doped, so that there are two *p*-type regions separated by an extremely thin *n*-type region. A simple wire connection is made to each region.

Fig. 40 (iii) is closer to the actual scale of the transistor, the shading representing the electrical connections of the emitter and collector leads to the flat faces of a very thin rectangular wafer of semiconductor. The collector contact is large because this is normally the output circuit of the transistor amplifier and will therefore be carrying more power than the input circuit, which usually contains the emitter. The area of the wafer may be anything from a fraction to several square millimetres, depending on the amount of power the transistor is required to handle. The faces of the wafer are *p*-type, and the *n*-type base region is an extremely thin layer separating them. The whole assembly will be enclosed in some sort of capsule to prevent mechanical damage and to eliminate light, which is similar to heat in producing intrinsic semiconduction effects. If it is a large power transistor, there will probably be some sort of metal heat sink associated with the capsule so that heat is removed from the semiconductor, particularly at the collector junction, and dissipated to the surroundings.

Let us now look in some detail at what takes place at the junction between the emitter and the base region. This is commonly called the emitter-base junction, or simply the emitter junction. Similarly, the junction between collector and base is often called the collector junction.

When a voltage in the forward direction is applied to an ordinary *p–n* junction diode, as described on page 74, the current consists of similar numbers of holes passing from *p* to *n* and of electrons passing from *n* to *p*. This is because the amounts of impurity in the *p* and *n* materials are similar and, hence, so are the respective numbers of holes and electrons in them.

If, however, a *p–n* junction is made where the amount of impurity in the *p*-type greatly exceeds that in the *n*-type material, the application of a voltage in the forward direction will give a current in the normal way; but this current will consist predominantly of holes passing over the junction from *p* to *n*, with a relatively negligible number of electrons passing in the opposite direction. This is the type of junction which is the emitter-base junction of the *p–n–p* transistor.

If for the moment we ignore the presence of the

collector junction, then, when the emitter junction is biased in the forward direction (i.e. a forward voltage is applied to it), a steady current of holes flows over the junction from the emitter region into the base region and out of the base terminal.

Under such circumstances, *X* in Fig. 41 is a typical hole path. The hole wanders at random in the base region, but there will be a tendency for diffusion to be towards the base terminal, where the concentration of holes is continually being reduced by the removal or neutralisation of holes by the negative pole of the battery which is connected to this point.

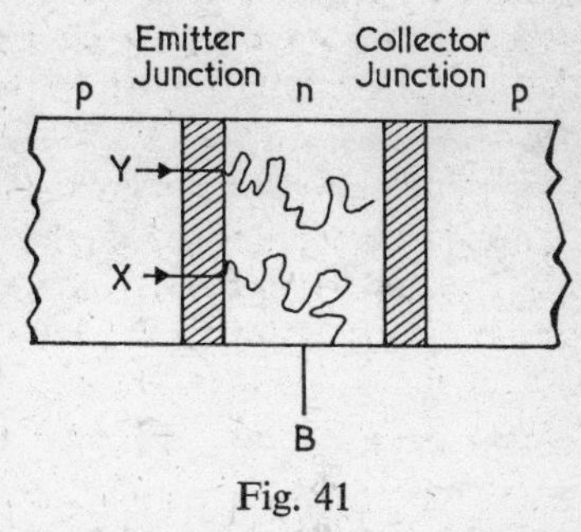

Fig. 41

Before considering the effect of the presence of the collector junction, let us review the situation in a *p–n* junction when a voltage is applied in the reverse direction and no current flows.

In Fig. 42 (i) is shown the equilibrium state of affairs in a *p–n* junction with no external voltage applied, the potential hill being formed in the manner discussed on page 69. In Fig. 42 (ii) is shown the same junction biased in the reverse direction by an external battery. No electrons diffuse from right to left because the most energetic electron in the *n*-type material (*A*) has insufficient energy to surmount the hill if it arrives at the depletion layer. Similarly, no holes will pass from left to right. There is no question of electrons moving from left to right across the depletion layer (or holes moving from right to left) because there are no free electrons available in the *p*-type material (or free holes in the *n*-type).

If free electrons could in some way be made available in the *p*-type material at the depletion layer (or free holes in the *n*-type), they would at once be swept through the junction by the potential hill, which is in the correct

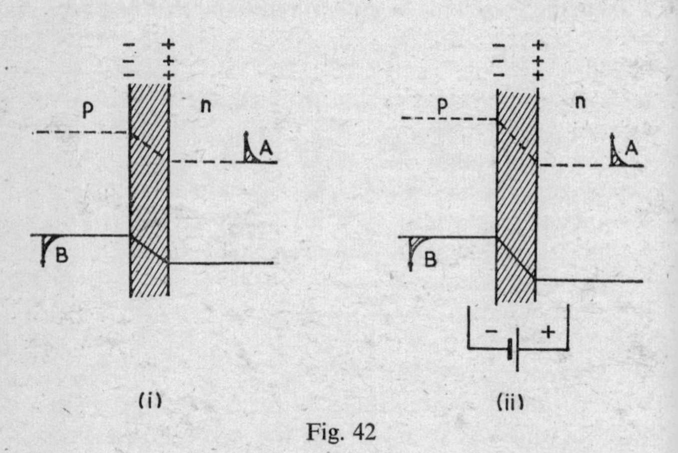

Fig. 42

direction for such minority carriers. (Minority carriers are holes present for some reason in *n*-type material or electrons in *p*-type.)

In Fig. 43 (i) and (ii) a *p–n–p* transistor is shown connected to batteries so that the emitter is positive with respect to the base and the collector is negative with respect to the base. The emitter junction is thus biased in the forward direction, and the collector junction is biased in the reverse direction. The current across the forward biased emitter junction will consist predominantly of holes, as shown.

The situation in the base region is shown in greater

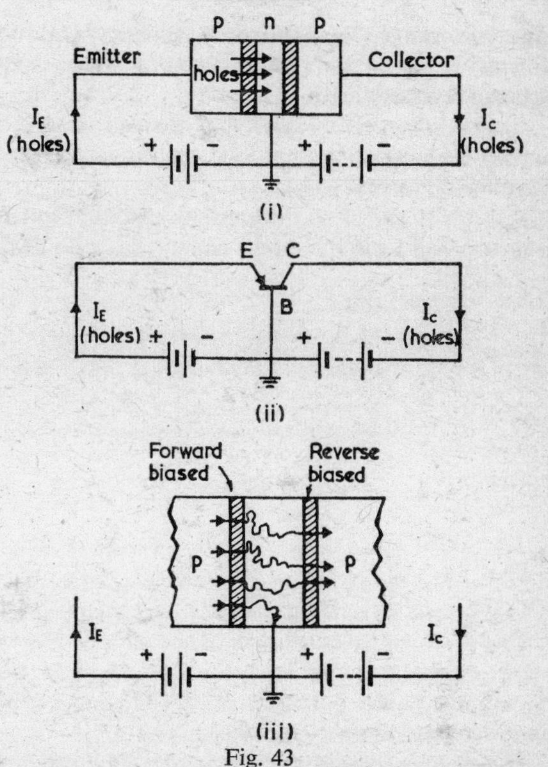

Fig. 43

detail in Fig. 43 (iii). Holes cross the emitter junction and, once in the base region, tend to drift at random. Because of the geometry of the base region, many more of the randomly moving holes will reach the large area collector junction than will reach the small area base contact. Any hole reaching the reverse biased collector junction is at once swept through the depletion layer into the collector region and thence to the collector terminal.

A large percentage of the holes which make up the

emitter current (I_E) find their way to the collector junction and a current (I_C) flows in the collector circuit, in spite of it being reverse biased.

In particular, it is found that a small change in the emitter current (ΔI_E) will produce a corresponding change in collector current (ΔI_C).

$\Delta I_C/\Delta I_E$ is called α, the current amplification of the transistor, and α is commonly as large as 0·98. A change

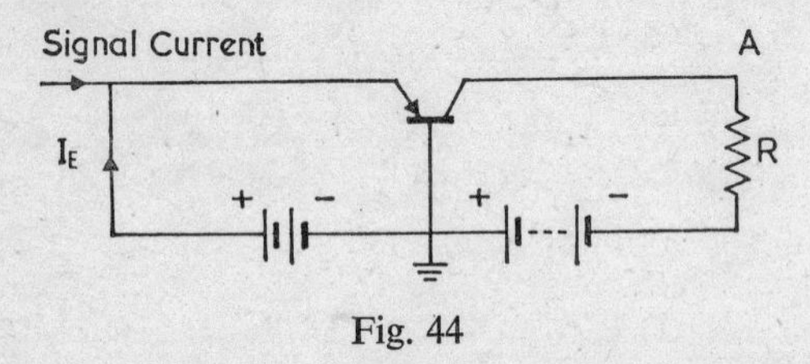

Fig. 44

of current of 1 mA in the emitter circuit will produce a change of very nearly 1 mA in the collector circuit.

In Fig. 44 a resistive load *R* has been placed in the collector circuit. Any change in the emitter current will produce a change in collector current, and hence a fluctuation in the potential at *A*. (Note the similarity with the valve discussion on page 81.)

The change in emitter current might be produced by adding a signal current to the steady current I_E, as shown. The resistive load *R* can be fairly large without upsetting the transistor action since it is placed in series with the collector junction, which has extremely high resistance because it is reverse biased.

A junction diode biased in the forward direction will constitute a low resistance in any circuit because a small voltage, or change in voltage, will cause a considerable current, or change in current, in the forward direction (volts/amperes = ohms). Conversely, a junction diode biased in the reverse direction will constitute an extremely high resistance since only a minute current (theoretically zero) will flow and any change in the voltage in the reverse direction will hardly produce a significant change in current.

Applying these considerations to a properly biased junction transistor, the emitter circuit will have a low resistance and the collector circuit a very high resistance.

Typically, a signal voltage of, say, 0·1 volt might be arranged to produce a change in emitter current of 1 mA. The change in collector current would also be approximately 1 mA, and the load resistor through which it flowed might be 5 kilohms. The change in potential at the point A would then be 1 mA × 5000 ohms = 5 volts.

Thus a change in voltage of 0·1 volt on the input side will cause a change in voltage of 5 volts on the output side. The voltage amplification, or stage gain, is 50.

As with the valve amplifier, the circuit components and the battery or supply voltages to carry out a particular amplification function must be chosen to suit the transistor which is to be used. The detailed properties and behaviour of a valve or transistor are best summed up in a set of curves called 'characteristics', which are discussed in the next chapter. The use of these characteristics to design an amplifier is discussed in Chapter 6.

n–p–n Transistor

The junction transistor described above consists of two junction diodes made in the same piece of solid material so that the current in one diode controls the current in the other. The transistor described was a *p–n–p* arrangement with its operation dependent upon holes, injected from the emitter into the base region, which diffuse across to the collector.

Transistor action can also be obtained in the complementary arrangement, consisting of an *n–p–n* transistor with the operation dependent upon electrons injected from the emitter into the base region which pass thence to the collector.

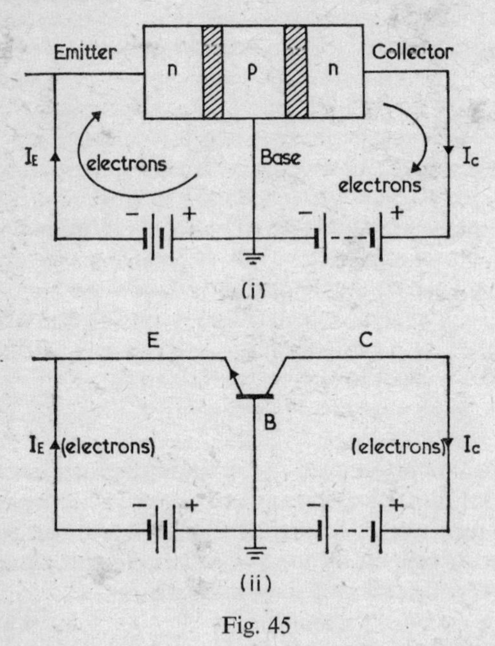

Fig. 45

The symbol for the *n–p–n* transistor (see Fig. 45) has the arrow on the emitter pointing in the opposite direction to that in the *p–n–p* transistor (in both cases the arrow shows the direction of flow of conventional current, i.e. positive charge). Note that the supply voltages for the emitter and collector circuits are opposite in polarity to those in the *p–n–p* transistor.

CHAPTER 5

Characteristics

The complete electrical behaviour of any valve or transistor can be described by stating the interrelation of the currents and the voltages between all the electrodes. These relationships can conveniently be displayed graphically, and the various curves are known as the 'characteristics' of the device. In principle, all the characteristics should be available to the designer proposing to use the device in a circuit. But, in practice, some of the characteristics of any device have little influence on performance, and often one of the characteristics is all that is required to enable a circuit to be designed and put together for laboratory test.

Diodes

The most important property of the diode, thermionic or solid state, is that it conducts current in one direction only. In the ideal case, a voltage applied one way round will give no current through the diode, but if the voltage is reversed, there will be a flow of current which increases with the size of the applied voltage.

There may be cases where we are interested in more than the fact that the diode is unidirectional. We may wish to know the way in which the forward current depends upon the forward voltage. This relationship is quite complex, mathematically, and most problems which require such information are tackled graphically, using an experimentally determined graph of I against V for the diode. This is called the diode characteristic.

The way in which the full I–V relationship arises in the thermionic and the junction diode has been discussed in Chapter 3, and the theoretical curves are shown on pages 65 and 75 respectively.

Bearing in mind that the thermionic diode is normally operated in the space-charge limited region, Fig. 46 can be taken to represent, for practical purposes, the characteristic of either the thermionic or the junction diode. The general shape will be the same for all diodes, but the current and voltage scales will differ. In particular, the bigger the diode—junction or thermionic—the bigger, usually, will be the current for a given voltage.

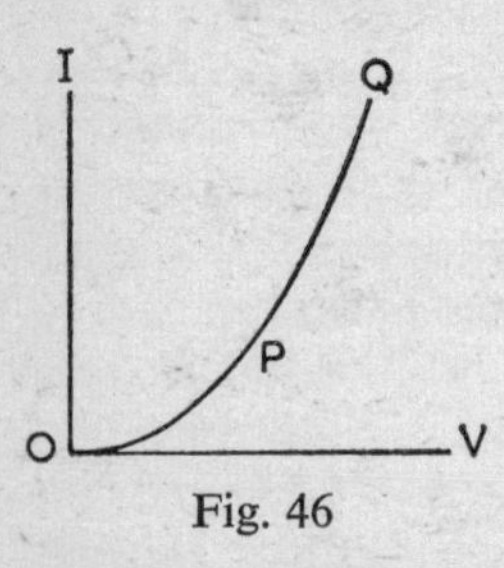

Fig. 46

It should be remembered that the whole of OPQ is a curved line. But a region like OP will often be referred to as the 'curved' part of the characteristic and PQ as the 'straight' part. When such terms are used they are only relative.

A characteristic such as Fig. 46 describes the behaviour of the diode alone. In practice, we shall be much more concerned with the behaviour of a circuit containing this diode and some sort of load, which we can represent by a resistance R.

The diode shown in Fig. 47 (i) is a thermionic valve, but the same argument applies to a junction diode, illustrated in Fig. 47 (ii). The battery voltage is V_B, the voltage across the diode is V and that across the resistor is V_R. The current in the circuit is I.

$$V_B = V + V_R = V + IR$$
$$V = V_B - IR.$$

There are two variables in the equation: V and I, V_B and R being constant in the circuit described. This equation can thus be represented as a straight line on a

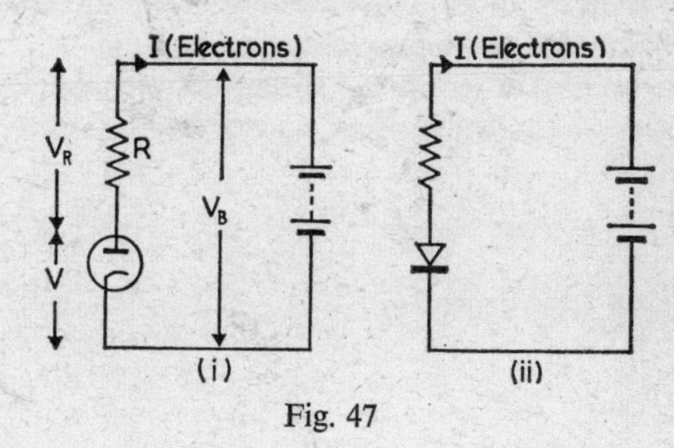

Fig. 47

graph of I against V. In particular, it can be shown as a straight line on a graph which also shows the characteristic of the diode used in the circuit (see Fig. 48 below).

OPQ is the characteristic of the diode and *LM* is the line representing the equation $V = V_B - IR$. This line is called the load line for resistance R.

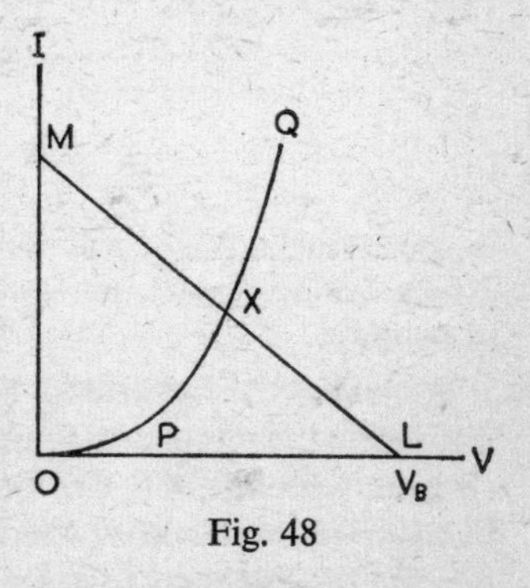

Fig. 48

The load line is drawn by fixing the point L on the voltage axis at a voltage equal to V_B (which is the value of V in the equation when $I = 0$) and by fixing the other end M on the current axis at a value $I = V_B/R$ (which is the value of I when $V = 0$).

The current and voltage conditions in the circuit with

load R and battery V_B must always be represented by some point on LM.

But the current and voltage conditions in the particular diode used in the circuit must always be represented by some point on the diode characteristic OPQ.

The point X where the load line and the characteristic intersect is the only point which satisfies both these

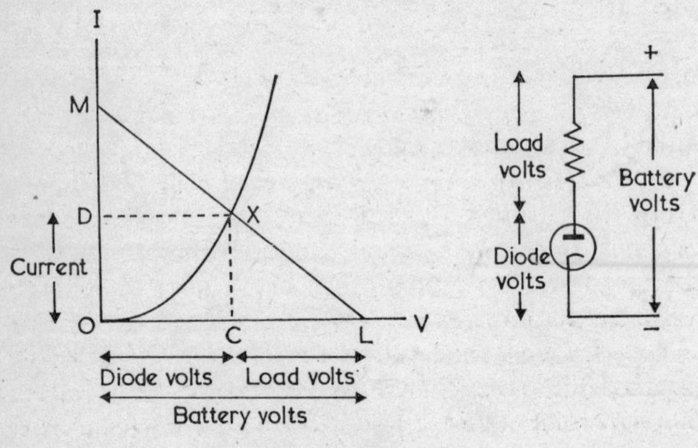

Fig. 49

requirements. This will describe the actual state of affairs in the circuit, and it is called the working point of the circuit.

With X as the working point (Fig. 49), the current flowing in the circuit, and in particular through the diode and the load resistor, will be given by OD. The corresponding voltage across the diode will be given by OC, and the voltage across the load due to the current flowing in it will be given by CL. The battery voltage is represented by OL.

On page 96 is reproduced the characteristic of a diode used in the circuit of Fig. 49.

On this characteristic construct the load line for a resistor of 2000 ohms and a 50-volt battery. Hence, determine the current flowing in the circuit, and the voltages across the diode and the load resistor.

Suppose the load resistor were changed to 5000 ohms. Draw the new load line, and determine what would now be the current in the circuit and the voltage across the load.

Repeat for a load resistor of 1250 ohms.

Suppose that the load were changed back to the original 2000 ohms, but the supply voltage changed from 50 to 20 volts. Draw the new load line.

(*See diagram overleaf*)

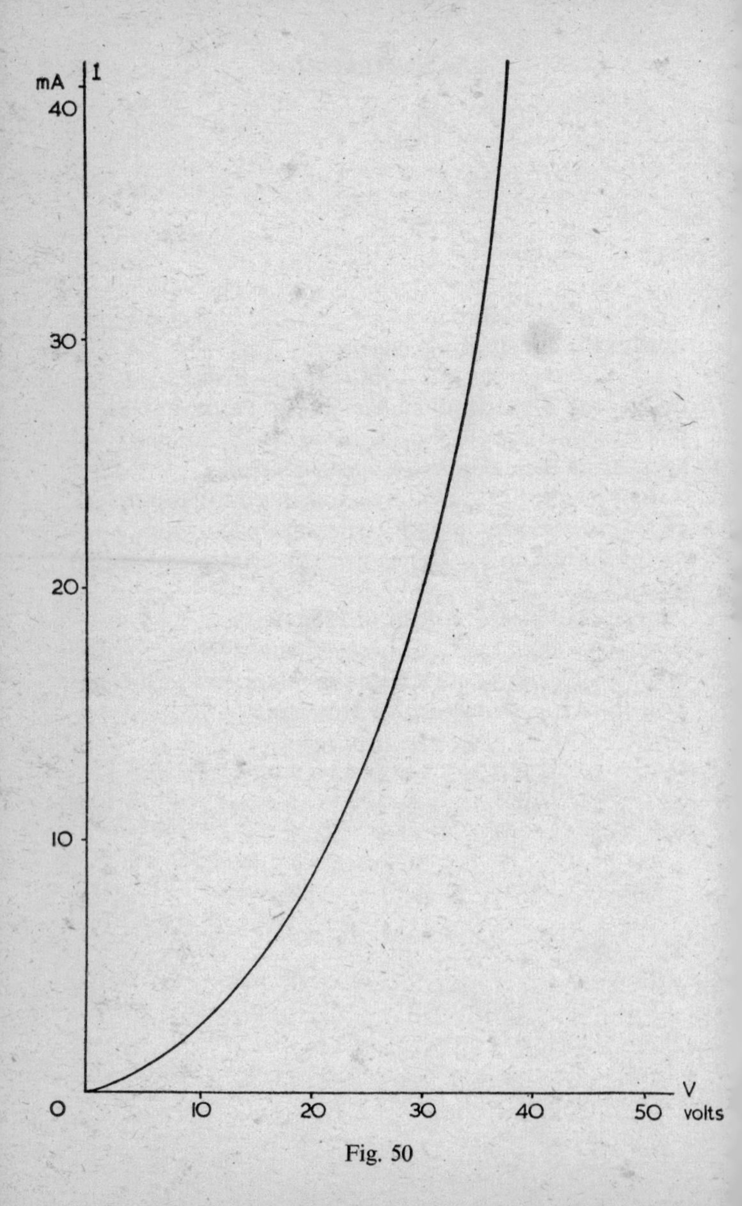

Fig. 50

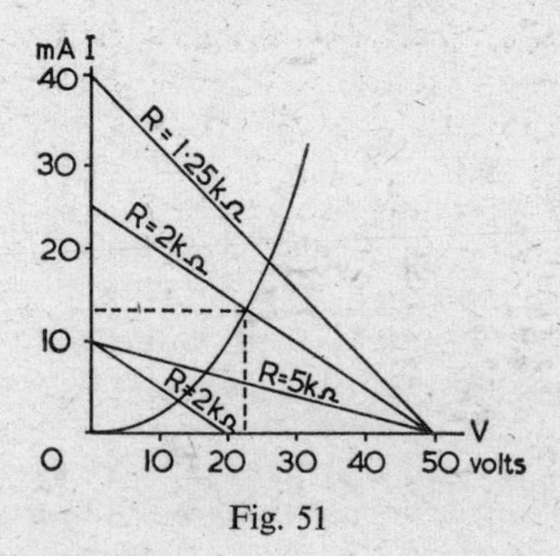

Fig. 51

Note the general conclusions: that the smaller the load resistance, the more nearly vertical the load line, and that load lines drawn for the same resistance but different supply voltages are parallel.

V_B	R	I	V_D	V_R
50	2000	13	24	26
50	5000	6.4	18	32
50	1250	17.6	28	22
20	2000	4	12	8

(*from Fig. 50*)

On page 99 is shown a diode characteristic *OT*. On the same diagram, the straight line *OD* represents an 'ideal' diode characteristic.

Assuming the same simple circuit as before—i.e. diode, resistor and supply voltage in series—draw the load line for a resistor of 2500 ohms and a supply voltage of 15 volts. Mark the resultant working point on the 'ideal' characteristic *OD* as *C*, and the 'real' characteristic *OT* as *S*.

Now draw two other load lines for the same resistor but for supply voltages of 10 and 5 volts. Mark the working points on the ideal characteristic as *B* and *A*, and on the real characteristic as *Q* and *P* respectively.

Hence, discover the change in current produced by a change in supply voltage from 15 to 10 volts and from 10 to 5 volts (*a*) with the ideal characteristic and (*b*) with the real, curved characteristic.

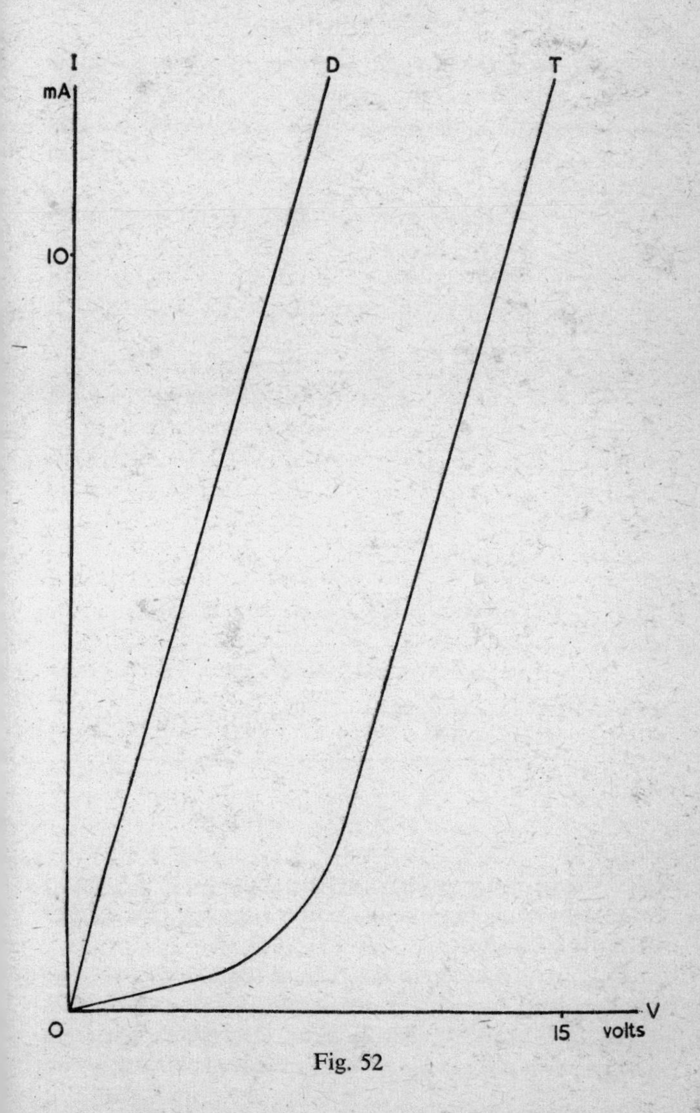

Fig. 52

With the ideal straight line characteristic, the same change in current is produced by any change of 5 volts in the supply, e.g. the change from 15 to 10 volts and that from 10 to 5 volts both produce the same change in circuit current.

But the real characteristic is curved, and consequently

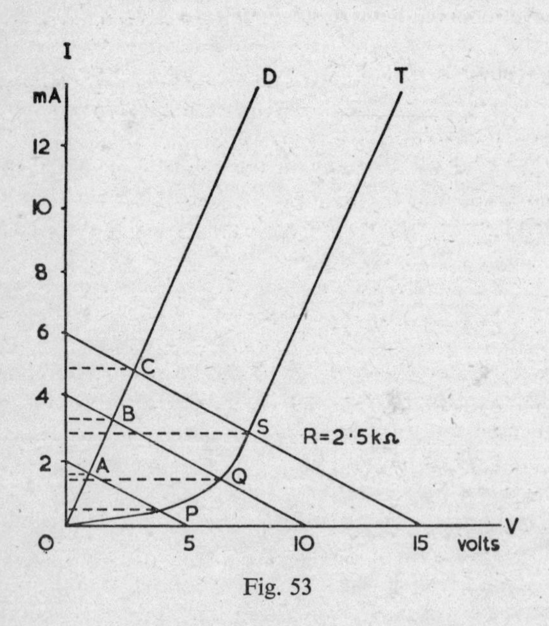

Fig. 53

the change in supply voltage from 15 to 10 volts will produce a much bigger change in current than the change from 10 to 5 volts.

This non-linearity is often important in the case of diodes, and also with triodes, pentodes and transistors (see pages 103–110). Occasionally, it produces desirable effects, such as the ability to mix two signals of different frequency and extract the difference frequency, but more often it is the cause of undesirable distortion.

Suppose, for instance, that the supply voltage was not taken from a battery, but from a source which gave a sinusoidal variation with time between 15 and 5 volts, as shown in Fig. 54 (i).

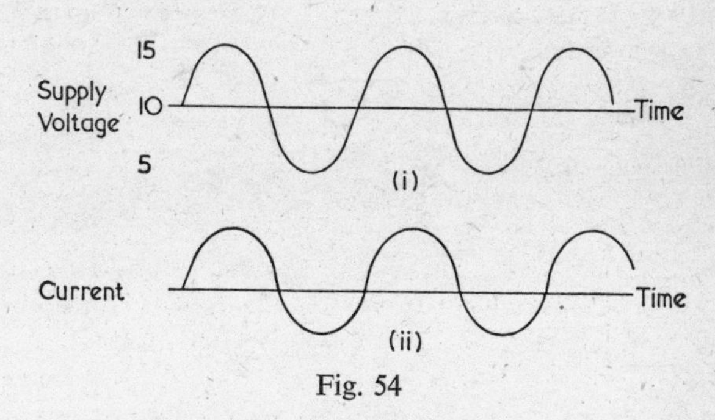

Fig. 54

Using the 2500 ohms load line and the curved characteristic of the last example (Fig. 53), the working point will move up and down the characteristic between the points P, Q and S, and the resultant current in the circuit will be like Fig. 54 (ii).

The upper half-cycles of current will be approximately half-sine waves and will certainly be larger than the lower half-cycles, which will be squashed. The current waveform is a distorted version of the supply voltage waveform.

If it can be arranged that the working point remains on the reasonably straight part of the characteristic, say above S, then the distortion is much reduced. In the example just examined, if the load were reduced from 2500 to 1500 ohms and the supply variation confined to between 15 and 10 volts, there would be a very much less distorted current waveform.

Thermionic Triode

In the triode valve, the current depends upon the voltage between anode and cathode (as in the diode valve), but it also depends upon the voltage between the grid and the cathode. The diode behaviour is shown as a characteristic with two axes, I_a and V_a. The behaviour of the triode could be represented by adding a third axis for V_g, thus producing a three-dimensional valve characteristic.

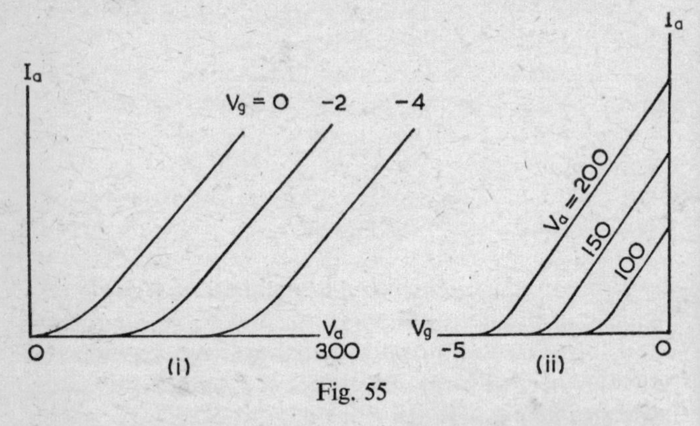

Fig. 55

Such a characteristic is an interesting solid shape, but is difficult to manipulate and to understand. Instead, the properties of the triode valve are normally displayed on a pair of two-dimensional characteristics (see Fig. 55).

The voltages V_a and V_g are relative to the cathode of the valve, which is usually held at earth potential, i.e. zero, when the characteristics are being determined.

In Fig. 55 (i) is shown the way in which I_a varies with V_a for various fixed values of V_g. The grid voltage in a triode is not normally allowed to go significantly positive with respect to the cathode because it is so close to the cathode that it would attract a very large fraction of the electron current which is meant to cross to the anode.

Notice that, when the grid is negative, the anode may be substantially positive without current crossing to the anode. For example, the $V_g = -4$ characteristic cuts the voltage axis on the I_a–V_a graph at over 100 volts, and at this point $I_a = 0$ and the valve is said to be 'cut off'.

In Fig. 55 (ii) is shown the way in which I_a varies with V_g for various fixed values of V_a in the same triode valve.

Note that each diagram contains a family of curves, instead of just one, and that therefore a large amount of the information on one set of characteristics is also contained on the other.

In general, we shall use the I_a–V_a characteristics more than the others.

Both sets of characteristics have regions where they are nearly straight and parallel, and other regions where they are curved. In an amplifier, the circuit is normally designed so that the valve operates in the straight and parallel region. Under these circumstances, it is the slope of the characteristics which is important and, since this is the same for the whole family of curves in this region, it can be quoted simply as a number. Such a number quoted for a particular valve is called a 'valve constant'.

The slope of the I_a–V_a characteristic is called the anode slope conductance, $g_a = \dfrac{\Delta I_a}{\Delta V_a}$, for a constant value of V_g.

$g_a = \dfrac{1}{r_a}$, where r_a is called the anode slope resistance, or a.c. resistance, and is measured in ohms.

The slope of the I_a–V_g characteristic is called the mutual conductance, $g_m = \dfrac{\Delta I_a}{\Delta V_g}$, for a constant value of V_a, and it is usually measured in milliamps per volt.

Triode with Load Line

If the grid voltage of a triode changes, the current through the valve will change. In Fig. 56 a resistive load is included in the circuit. We have seen (page 81) that a signal voltage applied to the circuit so as to cause a variation in V_g will cause a variation in I_a, and hence in the potential drop across R. The change in potential at A may be many times greater than the change in potential on the grid due to the signal voltage.

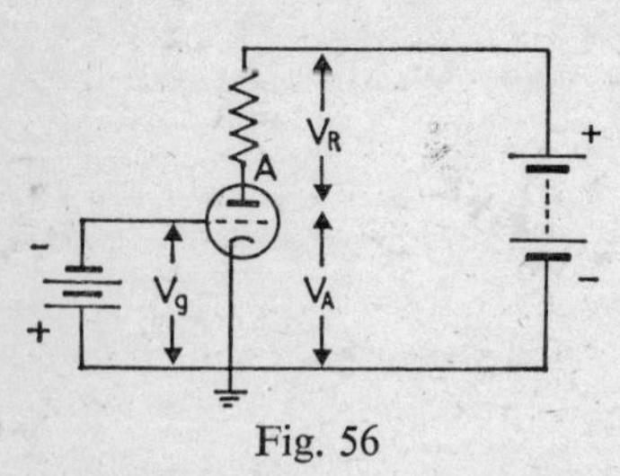

Fig. 56

Amplification is thus achieved and the maximum theoretical value is μ, the amplification factor of the valve itself. The actual amplification obtained depends upon R, and we shall now see how to use the triode I_a–V_a characteristic with a load line in order to choose the most suitable value for the load resistor under various circumstances.

In Fig. 57 BC is the load line, for a resistance of 10 000 ohms and a battery voltage of 250, drawn on the I_a–V_a characteristics of a triode valve. It is drawn in the same way as the load line on the diode characteristic (see page 93). One end of the line is fixed by the battery voltage and the other end by the battery voltage divided by the resistance (250/10 kΩ = 25 mA).

If the current axis does not extend to sufficiently high values to allow the point C to be marked, the load line can be drawn by fixing B as before and then drawing a line of the appropriate slope, i.e. so that the voltage increment of the line divided by the corresponding current increment gives the correct resistance. This would be the way to draw a 1000 ohms load line on the above characteristic.

Referring again to the 10 kΩ load line, the points *P*, *Q* and *R*, where it intersects the 0, —2 and —4 characteristics, give the conditions in the circuit for these three values of grid voltage.

If the grid voltage is made —4, then the working point

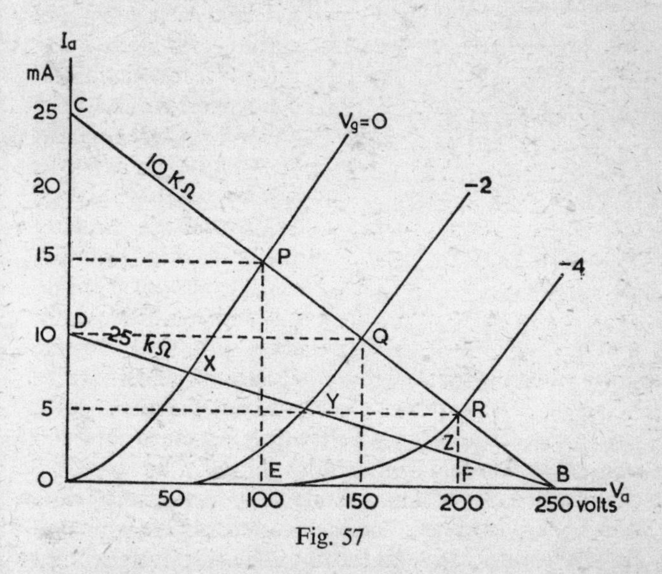

Fig. 57

is *R*. There will be a current of 5 mA flowing in the circuit, and the potential at the point *A* where the resistor joins the anode will be 200 volts (*F*). If the grid voltage is changed to —2, the working point becomes *Q*, and if the grid voltage is made 0 the working point will be *P*, with a current of 15 mA and a potential at the anode of 100 volts (*E*).

A change in the grid potential of 4 volts gives a change

in current of 10 mA and a change in the anode potential of 100 volts. The amplification is 25.

If the grid potential is made to vary continuously between 0 and −4 volts, then the working point will move continuously up and down the load line between *P* and *R*. If the variation is made sinusoidal with time by applying an alternating signal voltage to the grid, then the current will vary sinusoidally between 15 and 5 mA, and the anode voltage will vary sinusoidally between 100 and 200 volts (see Fig. 58).

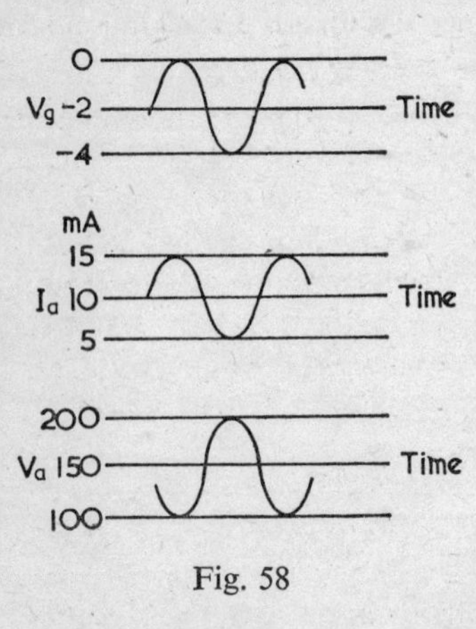

Fig. 58

The anode voltage variation, which will become the output signal in an amplifier, is greater than the grid voltage variation, which will be the input signal. Both are sinusoidal in form, with the anode voltage variation in antiphase to that of the grid.

Because both voltage waveforms are sinusoidal, the amplification is said to be undistorted. There is no distortion because the excursions of the working point up and down the load line never take it outside the straight region of the characteristics. Anywhere within this region a given change of grid voltage produces the same change in current.

This would not have been the case if the load resistor had been 25 000 ohms, and the grid potential had still varied between 0 and −4 volts. The load line would now have been *BD*, and the excursions of the working point would have been between *X* and *Z*, where *Z* definitely lies in the curved region.

Under these circumstances, the change in grid voltage from −4 to −2 would give a smaller change in current than the equal grid voltage change from −2 to 0. There would be a correspondingly smaller change in anode voltage, and the half-cycle of output voltage due to the movement of the working point between *Y* and *Z* would be smaller than that due to the movement between *Y* and *X*. The amplification will be greater than it was with a 10 000 ohm load, but the output will be a distorted version of the input.

The greater the curvature of the characteristics, the greater will be the distortion. In certain amplifier applications, more distortion can be tolerated than in others and the circuits will be designed accordingly. Even the 'straight' portions of the characteristics are, in fact, slightly curved, so there is always some measure of distortion.

In the example with the 25 kΩ load just quoted, a signal giving a 4-volt fluctuation in grid potential results in distortion. If, however, we wished to amplify a fluctuating signal that was only 2 volts in total excursion instead of 4 volts, it could be accomplished without significant distortion by causing it to swing the grid potential from 0 to −2 volts. The working point would now move between *X* and *Y*, always in the straight region.

Pentode Characteristics with Load Line

In Chapter 4 we have seen that a pentode valve is essentially a triode with two additional grids to improve performance. Generally speaking, the extra electrodes are maintained at fixed voltages and the pentode action is the same as the triode action—i.e. fluctuating voltages applied to the control grid give fluctuations in valve current and, hence, amplified voltage fluctuations at the anode.

Although the I_a–V_a characteristics for the pentode are very different in shape from those of the triode, the load line construction is employed in exactly the same way to decide how the circuit will behave when there is a resistive load in series with the pentode.

The pentode I_a–V_a characteristics shown in Fig. 59

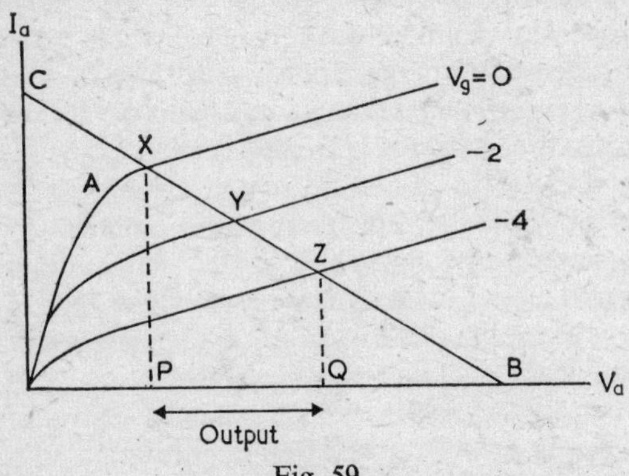

Fig. 59

have a 'curved' and a 'straight' region, with the straight portions of the set much more nearly parallel to the voltage axis than in the triode. The load line *BC*, shown above, cuts the characteristics in the straight region, but a bigger resistance would have given a line cutting the $V_g = 0$ characteristic in the curved portion on or below the shoulder of the curve *A*.

Typically, an amplifier using a pentode valve might have an input signal voltage which made the grid potential swing between 0 and − 4 volts. The working point would then move to and fro along the line *XYZ*, and the anode potential would swing between the values given by *P* and *Q*. Thus the input voltage swing would be 4 volts and the output *PQ* might be perhaps 150 volts.

Note that the slope of the pentode I_a–V_a characteristic in the straight region is such that the r_a of the valve $\left(\frac{\Delta V_a}{\Delta I_a}\text{ for a given } V_g\right)$ is extremely high compared with that of the triode.

Transistor Characteristics and Load Line

The transistor characteristic which is used for the load line construction is very similar to the above pentode

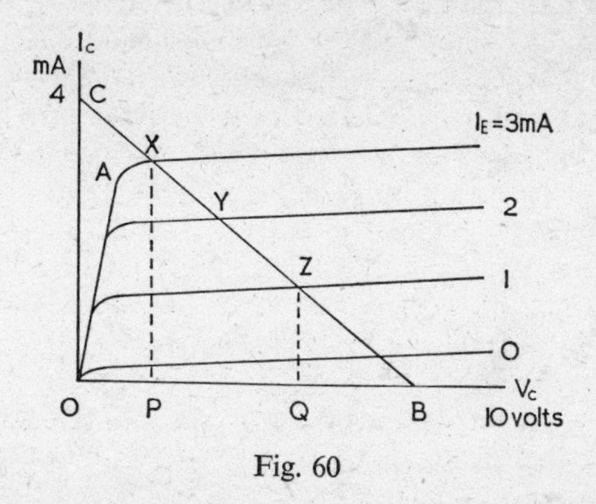

Fig. 60

characteristic, but with a rather more satisfactory shape for amplification purposes since there is no long sloping shoulder.

Fig. 60 is an I_C–V_C characteristic for various values of I_E, i.e. it shows the way in which the collector current is affected by the collector voltage for various fixed values of emitter current.

In the normal working region of the transistor, where the characteristics are straight and parallel, the collector voltage has comparatively little influence on the collector current, which is largely determined by the emitter current.

The collector and emitter currents are very nearly equal, and a change in emitter current of 1 mA will cause a change in collector current of almost 1 mA, i.e. the current amplification factor is nearly unity ($\alpha = 0{\cdot}98$).

There is a small collector current even when $I_E = 0$. The collector junction is reverse biased, and the current flowing when $I_E = 0$ is just that which would be expected in a reverse biased junction diode (see page 75).

The load line BC is drawn in the ordinary way, i.e. the point B is the battery or supply voltage and the point C is the battery voltage divided by R, the load resistor connected in series between the collector and the battery. If I_E is caused to fluctuate between 1 and 3 mA by the application of a signal current, then I_C will also fluctuate, and the collector of the transistor where it is connected to the load R will fluctuate in potential by PQ volts.

The fluctuation in signal current will be caused by a small signal voltage—much smaller than PQ—and hence there will be voltage gain in the amplifier.

As long as the working point remains in the straight region of the characteristics, there will be negligible distortion. If, however, too large a load resistor is used, the load line will cut the characteristics in the region OA and distortion will occur. But even with such a load line, distortionless amplification can be obtained, provided that the signal to be handled is small and only causes the working point to move up and down the load line in the straight region of the characteristics.

Other Characteristics

At the beginning of this chapter it was stated that, although several sets of characteristic curves were required to describe fully the behaviour of a valve or transistor, a single set might suffice for the first design of an electronic circuit. The I_a–V_a or the I_C–V_C characteristic was used to show the behaviour of a circuit with a load in series with the valve or transistor, and the discussion was directed towards the question of amplification, which is to be dealt with more fully in Chapter 6.

It was also seen (page 102) that the information contained in a set of I_a–V_a characteristics could alternatively be presented as a set of I_a–V_g characteristics. Similarly, the I_C–V_C curves plotted for various fixed values of I_E could be presented as a set of I_C–I_E curves for fixed values of V_C. Such a set of curves—nearly straight lines—is sometimes known as the forward transfer characteristics of the transistor. Occasionally, in dealing with transistors, another characteristic may be useful. This is sometimes called the input characteristic and is a graph of the emitter current against emitter voltage for fixed values of collector voltage. This will, for instance, allow the effect of a signal voltage applied between emitter and base to be assessed in terms of the emitter current produced, and hence the consequent collector current and the fluctuation in voltage across the load.

The principal characteristics used, however, will be those discussed in this chapter, and their use in conjunction with a load line will be relevant to the subsequent discussions of amplification, switching and pulse generation. In some of the applications, the transistor will be used in a slightly different way from what we would expect after the discussion so far. Consequently, the characteristic used will be slightly different; but we shall find that all the ideas about load line, working point and distortion can be quickly and easily applied in the new circumstances.

CHAPTER 6

Amplifiers

The main purpose of an amplifier is to take in a signal from, say, an aerial, or a gramophone pick-up, or a photocell and to amplify it so that it can be used to drive some load like a loudspeaker or a motor.

A transformer will convert a small alternating voltage into a larger one, but only at the expense of a corresponding reduction in the associated current, so that the power on the output side is the same as that on the input side (in practice, there will be some power losses in the transformer). If, however, a signal voltage or current is to be magnified without a corresponding reduction in current or voltage respectively, then an active circuit element, such as a valve or a transistor, must be used so that overall power gain from input to output is achieved.

In earlier chapters we have seen that this can be done by making the electrical voltage or current in one circuit control the amount of current delivered by a battery or other power supply to the load in another circuit. The electrical signal applied to the grid, or emitter, circuit controls the current flow in the anode, or collector, circuit containing the load. In this chapter we shall look at amplifier circuits and see the reason for the other components that must be used in conjunction with the valve or the transistor in practical equipment.

Standard Amplifier Configurations

So far, in discussing the use of valves and transistors as amplifiers, the input circuit has been connected between grid and cathode, or between emitter and base, and the output circuit has been between anode and cathode, or collector and base. Thus in the valve amplifier the cathode has been taken as the electrode common to the input and output circuits, and in the transistor amplifier the base has been the common electrode.

Satisfactory amplification can also be obtained with the grid or the emitter used as the common electrode. The input is then applied to the cathode (i.e. between cathode and grid) or to the base (i.e. between base and emitter), and the output circuit is between anode and grid or between collector and emitter.

In principle, any of these configurations will give satisfactory amplification, but there are certain secondary advantages in choosing one rather than another. The valve amplifier generally used has the common cathode connection that we have met in all previous discussions of triode and pentode behaviour. But the normal transistor amplifier has the common emitter configuration which has just been mentioned for the first time.

We shall look briefly at the way the transistor behaves when used in the common emitter configuration. But it is not really necessary to worry about what is happening inside the valve or transistor when setting out to design a circuit. The appropriate characteristic curves provide all the information required, and no other details of valve or transistor behaviour need be considered until, perhaps, refinements in the original design are contemplated.

In the ordinary way, inside the transistor the emitter-base junction is biased in the forward direction, and there is an injection of charge carriers into the base region from the emitter (holes in the *p–n–p* transistor and electrons in the *n–p–n*). Nearly all these carriers cross the base region to the collector, and only a few reach the base terminal. In Fig. 61 (i), of 100 carriers entering by the emitter terminal, 98 leave by the collector terminal, and the current amplification factor from emitter to collector is 0·98. In Fig. 61 (ii) the same thing is still happening, but it can be seen that, with this circuit configuration, a signal applied between base and emitter so as to produce a certain change in I_B will cause a change

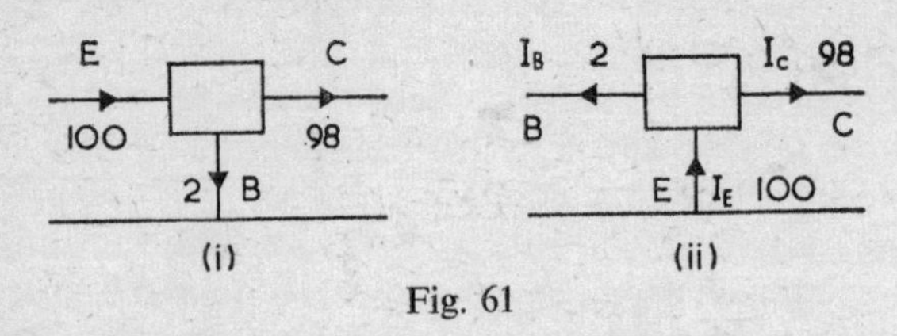

Fig. 61

in I_C, and that the current amplification factor between base and collector is about 50.

If a set of I_C–V_C characteristics is taken for a transistor in the common emitter state, they will have the same general appearance as those for the common base state, but the collector current scales are such that the change in collector current for a given change in I_B will be very much greater than for the same change in I_E.

We shall now design a common emitter transistor amplifier and a pentode valve amplifier using the appropriate characteristics (see Fig. 62 (i) and (ii)).

The transistor characteristic shows the way in which the collector current varies with collector voltage for various fixed values of base current. The valve characteristic shows the variation of anode current with anode voltage for various fixed values of control grid voltage.

In the valve, V_a will be positive with respect to the cathode and I_a will be a current of electrons flowing from cathode to anode. V_g will be negative with respect to the

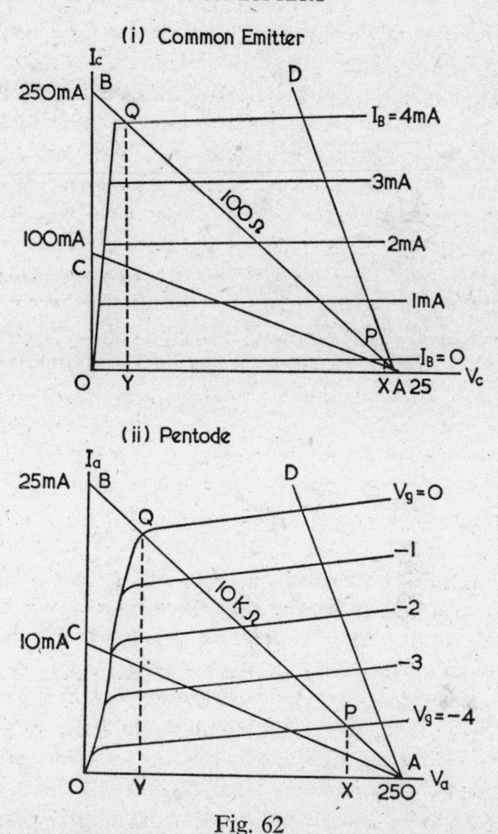

Fig. 62

cathode and the anode of the valve will be connected through the load to the positive side of the supply voltage.

The transistor used in the amplifier may be either a *p–n–p* or an *n–p–n* type. In either case, the appropriate charge carriers enter the transistor at the emitter and most of them leave at the collector, with a small fraction leaving at the base. I_E will consist of charge carriers entering the transistor, and I_C and I_B will be charge carriers leaving the transistor. In the *p–n–p* transistor these charge carriers will be holes, and in the *n–p–n* type they will be electrons. The *p–n–p* transistor will have its collector connected through the load to the negative terminal of the h.t. supply and the *n–p–n* collector will be connected through the load to the positive side of the supply voltage.

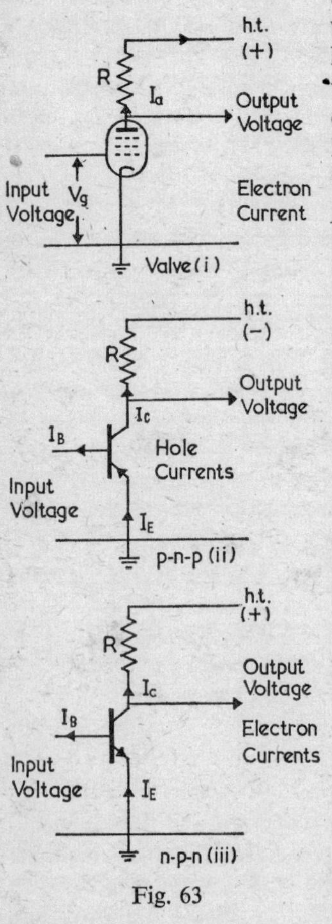

Fig. 63

In the valve amplifier (Fig. 63 (i)), when the size of the negative voltage applied to the grid increases, the amount of current flowing in the valve decreases and the potential drop across the load resistor decreases. The potential at the bottom of *R* will thus become closer to the h.t. value, which in the valve is positive. Thus, when the grid voltage becomes more negative, the output voltage becomes more positive. In this valve amplifier, the input and the output are in antiphase.

In the *p–n–p* amplifier (Fig. 63 (ii)), if the current I_B of holes leaving the base is increased, current I_C also increases and the potential at the bottom of R gets farther away from the h.t. value, which in this case is negative. So, if I_B increases in magnitude, the output voltage becomes less negative. To increase the current of holes leaving the base (I_B), the base voltage must be made more negative with respect to the emitter by the input voltage to the amplifier. Thus, if the input voltage becomes more negative, the output voltage becomes less negative. The input and output are in antiphase, as in the valve amplifier above.

Similarly, in the *n–p–n* amplifier (Fig. 63 (iii)), the input and output voltages are in antiphase, as in the two cases above. An increase in the electron current leaving the base (I_B) will cause an increase in the electron current through R and the output voltage will become less positive. To increase the electron current leaving the base, the base voltage must be made more positive with respect to the emitter by the input voltage to the amplifier. If the input voltage becomes more positive, the output voltage becomes less positive.

Suppose that the h.t. voltage available is 25 volts in the transistor amplifier and 250 volts in the valve amplifier. The load resistor is 100 ohms in the collector circuit of the transistor and 10 kilohms in the anode circuit of the valve.

This situation is represented on the characteristics (Fig. 62) by the load lines *AB*, which have slopes of 100 ohms and 10 kilohms respectively. The end *A* is on the voltage axis at a value equal to the h.t., and the end *B* is on the current axis at a value given by $\dfrac{\text{h.t. volts}}{R \text{ ohms}}$.

AC and *AD* are lines corresponding to loads of different resistance and will be considered later.

If the input to the transistor amplifier is such that $I_B = 0$, or the input to the valve amplifier makes the grid 4 volts negative with respect to the cathode, then the working point of the circuit is *P*. The voltage at the bottom of the load resistor *R*, i.e. at the end remote from the h.t. (Fig. 62), is given by the point *X* on the voltage axis.

If the input to the amplifier is now changed so that either $I_B = 4$ mA or $V_g = 0$, the new working point will be *Q*, and the new voltage at the bottom end of *R* will be given by *Y*.

Thus the 4 mA change in the base current or the 4 volts change in the grid potential causes a considerable change (*XY*) in the voltage at the bottom end of the load resistor. The voltage at the top end of this resistor is held at the h.t. value, and the potential difference across the load fluctuates about 20 volts in the transistor amplifier and about 150 volts in the valve amplifier, as the input signal causes the base current or the grid voltage to vary.

If the input signal varies with time in a certain pattern, then the way in which the potential at the bottom of the load resistor—the output voltage—varies with time can be found from the characteristics and the load line. For each value of I_B or V_g, the corresponding value of V_C or V_a can be found and plotted.

Using the load line *AB*, this construction has been carried out in Fig. 64 for one input with a square wave shape and for another with a sinusoidal variation with time.

Distortion

It is clear that, in the case to which Fig. 64 refers, the shape of the signal applied to the grid or the base is faithfully reproduced in amplified form in the load.

This will be the case as long as the load line cuts the characteristics at points where they are straight, parallel and equally spaced. This is the case with the load line *AB*, and would also be so with *AD*.

Under these circumstances, there is no need to plot corresponding points to confirm that the amplified signal in the load is an undistorted version of the input signal. The amplitude of the alternating voltage developed across the load can quickly be found by noting the points *X* and *Y*.

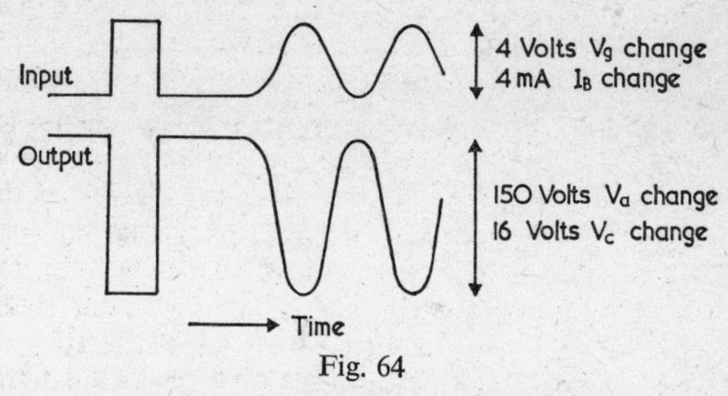

Fig. 64

If a load of greater resistance is chosen, giving a load line like *AC*, then a different situation arises. Imagine I_B changing from zero towards 4 mA and V_g changing from –4 volts towards zero. All is well as the working point goes up the load line from the first to the second characteristic. But, just short of the third characteristic, the load line runs into the curved region and there is no change of current in the transistor or the valve as I_B changes from about 1·8 to 4 mA, or V_g from –2·3 to 0 volts. Consequently, there will be no change in V_C or V_a over this range, and the signal appearing across the load resistor will be a grossly distorted version of the input signal.

Only when I_B is between 0 and 1·8 mA, or when V_g is between –4 and –2·3 volts, will undistorted amplification occur. For a sinusoidal variation of input with time (see Fig. 65), only the shaded portion of the signal will be amplified without distortion. The output voltage will have the distorted waveform shown.

Something less than half the input signal cycle is faithfully amplified and the remainder, shown dotted on the output curve below, will be clipped off. Such gross distortion

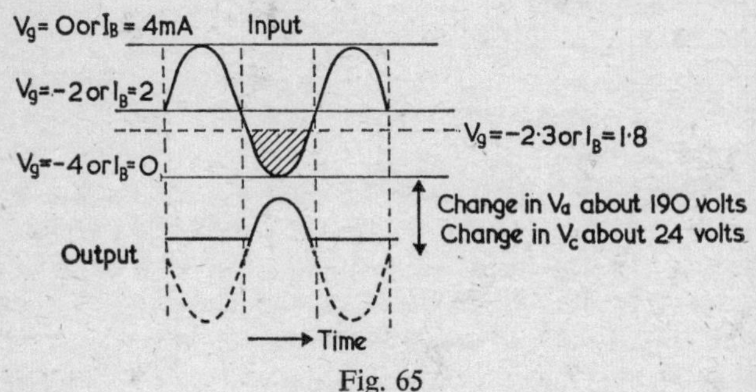

Fig. 65

of the shape of the signal is quite unacceptable in most amplifiers—in particular, those handling speech or music.

If a load of too great resistance is used with a particular transistor, distortion will occur. At first sight, it may seem that a low resistance should be used, giving a load line like *AD*. There would certainly be no question of distortion; but the voltage fluctuation obtained across the load, and hence the amplification, can be seen to be much smaller than before. The optimum resistance in the above example is something giving a load line like *AB*.

Bias

In starting to explain the design of an amplifier, it has been convenient to consider that the input signal causes I_B to vary between 0 and 4 mA, or V_g to vary between

–4 volts and 0. If the variation is sinusoidal, then the peak-to-peak value of the input is 4 mA or 4 volts.

But when such a signal is presented to a circuit for amplification it will normally be a genuine alternating signal. One of the two wires between which the signal voltage is developed will usually be at earth potential—often it is the earthed outer shield in a coaxial cable, which surrounds the live inner wire but is insulated from it. The potential of the live wire will fluctuate up and down about earth, or zero, potential. It will be a genuine alternating signal voltage which is presented to the input of the amplifier.

The available signal is one of 4 volts or 4 mA peak-to-peak amplitude, with one half-cycle positive and the other negative with respect to earth. What is required for the proper functioning of the amplifier is that V_g or I_B shall have this 4 volts or 4 mA peak-to-peak fluctuation, but that the basic condition for the operation of the valve or transistor shall be maintained, i.e. that the grid voltage of the valve is always negative with respect to the cathode, or that in the transistor the current I_B of charge carriers (holes in *p–n–p* or electrons in *n–p–n*) flows round the input circuit from emitter to base.

To ensure that the half-cycles of the signal which are in the 'wrong' direction do not take the amplifier out of its proper working region, a permanent input in the correct direction is provided in the circuit. This is called the bias voltage or current and, when the signal is added to it, the net input to the amplifier is always of the correct polarity. Thus the steady bias provided to a valve is a voltage making the grid negative with respect to the cathode. In the transistor amplifier, it will be a steady current I_B—of holes out of the base in the *p–n–p* case and of electrons out of the base in the *n–p–n* case.

In the above example, the steady bias must put the valve on the $V_g = -2$ volts characteristic, or the transistor on the $I_B = 2$ mA characteristic, before the signal is applied.

This bias voltage or current may be supplied to the amplifier by a second battery, or other d.c. supply, separate from the h.t. Such an arrangement is usually cumbersome and expensive. There are ways of providing the bias directly from the h.t. supply by inserting a suitable extra resistor in the circuit. These are described later in this chapter after a discussion of the factors which may influence the choice of bias voltage or current, and hence the steady working point of the circuit in the absence of any signal.

Choice of Steady Working Point

Let us consider again the characteristics for the valve and the transistor, and look once more at the way in which they may be used to settle the design of an amplifier. The characteristics are the same as were used earlier in this chapter, and the load line *AB* is again drawn. But now we shall take into account some of the problems which arise in the design of a real amplifier.

Looking again at the load line *AB* in Fig. 66, and remembering the discussion of bias in the last section, the working point *E* can be inserted on the diagram. This means that a steady bias of 2 mA for the transistor, or 2 volts for the valve, will be required.

Under these circumstances, a signal of 4 mA or 4 volts peak-to-peak can be amplified by the transistor and valve circuits respectively.

Suppose now that in a practical case the h.t. voltage available is less than that originally quoted. The bottom end of the load line will move from *A* to *K*. If the load

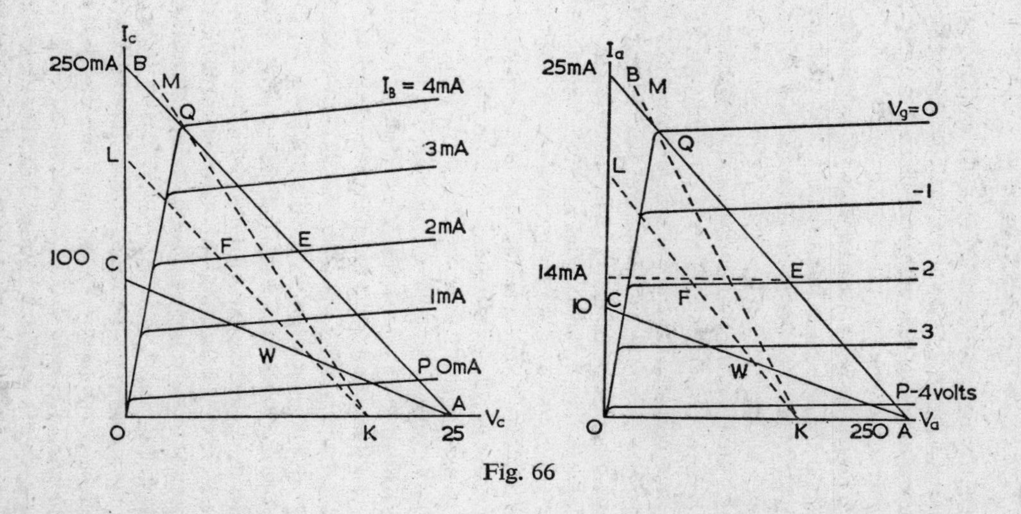

Ic
250mA
B
M
Q
L
100
C
F
E
W
O
K
25
A
Vc
IB = 4mA
3mA
2mA
1mA
P 0mA
Ia
25mA
B
M
Q
L
14mA
10
C
F
E
W
O
K
250
A
Va
Vg=0
-1
-2
-3
P-4volts

Fig. 66

resistor remains unchanged, the new load line will be *KL* parallel to *AB* and, with the previous values of bias, the new steady working point will be *F*.

With *F* as the working point instead of *E*, the amplifier will no longer be able to handle such large signals without distortion. When the signal takes I_B beyond the 3 mA characteristic or V_g beyond the –1 volt characteristic, distortion will occur.

There are now two courses open if distortionless amplification is to be achieved with the reduced value of h.t.

A new bias can be chosen which moves the steady working point *F* down the load line towards *K*. With a bias of $I_B = 1{\cdot}5$ mA or $V_g = -2{\cdot}5$ volts, it will be possible to accommodate signals of peak-to-peak value 3 mA or 3 volts without distortion. Thus the new amplifier, with reduced h.t., has the same gain as before because the load resistance is the same, but it cannot handle such large signals as it could do previously with the full h.t.

Alternatively, a new load line *KM* could have been drawn through *Q*. This would correspond to a lower load resistor than before, and hence the amplifier would have less gain. But if the bias is kept at $I_B = 2$ mA or $V_g = -2$ volts, the steady working point will be on the characteristic between *F* and *E*, and signals of 4 mA or 4 volts peak-to-peak can be handled as previously. The new amplifier with reduced h.t. can still amplify without distortion the same large signals as before, but the gain is reduced.

When first discussing the amplifier, the load line *AC* was dismissed as unsuitable because of the distortion introduced with a 4 mA or 4 volts peak-to-peak input signal.

However, it may be known that the signals to be amplified will always be smaller—say, 1 mA or 1 volt peak-to-peak. In this case, a steady working point, such as *W* on *AC*, would give distortionless amplification. The gain of the amplifier would be high because *AC* corresponds to a larger resistance than *AB*.

So far, the load line and the working point have been chosen in order to get the highest gain without introducing distortion when the largest expected input signal is amplified. In practice, certain other secondary considerations may also be involved.

There may be a requirement to keep the power con-

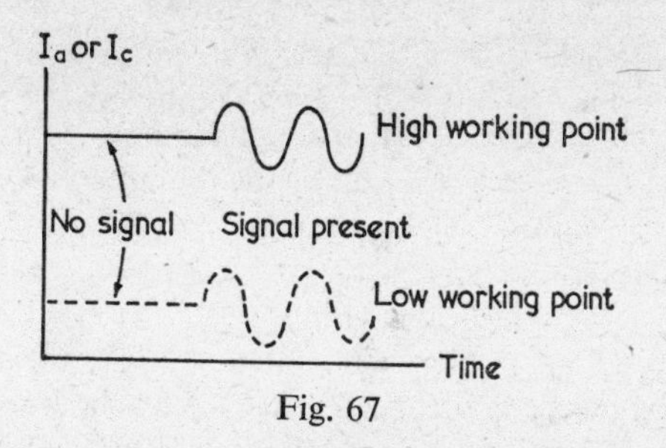

Fig. 67

sumed from the h.t. supply as small as possible. This could well be the case with portable equipment using batteries. If this is so, the designer will try to keep the working point as low as possible on the characteristics.

In the absence of any signal, the valve or transistor will pass a steady current I_a or I_C, which will be determined by the steady bias. When a sinusoidal signal is applied, I_a or I_C will vary sinusoidally about this steady value, as shown in Fig. 67. There is then a mixture of alternating and direct current or voltage at the output, just as the input consists of an alternating signal superimposed on a steady bias.

Only the a.c. component in the load is useful power, and the d.c. component of I_a or I_C is the average current drained continuously from the h.t. supply. A high working point like *E* (see Fig. 66) will give a large power drain from the h.t. supply, and a low working point like *W* means a smaller power consumption.

For a given a.c. power output, the lower the power consumed from the h.t. supply, the greater the amplifier efficiency will be and the less waste heat there will be to dissipate from the apparatus.

Spread and Tolerance in Components

The foregoing discussion has perhaps given the impression that load lines are drawn through a precise point on the characteristics and that the working point is chosen equally precisely to give exactly the required performance.

There would be little point in carrying out such an accurate construction on the characteristics published by the valve or transistor manufacturers. Somewhere in the makers' data sheets will be an indication of the spread in characteristics to be expected between different valves or transistors sold under the same type number. For ordinary commercial batches, this spread will often be large.

Even if this were not so, great accuracy would be unjustified because ordinary electronic components are made with about 10 per cent tolerance and are only readily available in certain preferred values. Thus, if calculation or an accurate construction shows that a resistor of 4850 ohms is required, the nearest available 'preferred value' will be 4700 ohms, and this may itself be wrong by 400 ohms.

When an amplifier is designed, care is taken to allow for these various uncertainties. Load lines and working points are deliberately chosen so that the likely errors due to these spreads and tolerances still leave the circuit working in the required manner. The final test of the design is to build the amplifier. If it does not have the required performance, it can be modified on the laboratory bench until it has.

Bias Circuits

Suppose we set out to achieve an amplifier represented by load line *AB* and working point *E* in Fig. 66. The load resistors (R = 100 and 10 000 ohms) are shown in the

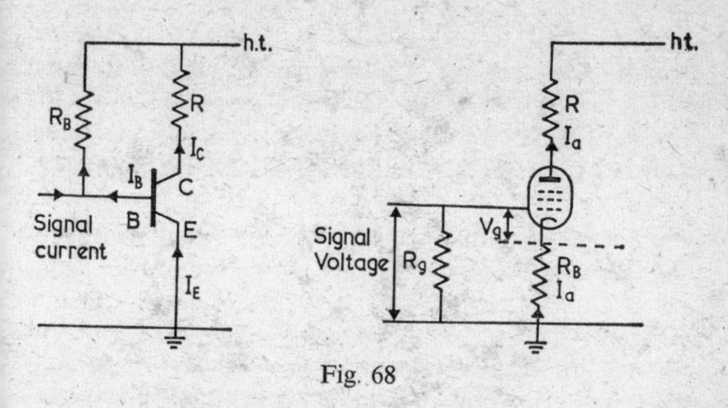

Fig. 68

circuits in Fig. 68 above. By the correct choice of R_B, the required steady bias is obtained directly from the h.t. without the need for a separate bias supply.

It is valuable to consider in more detail the conditions which will hold in the amplifier when the correct steady working point has been achieved and no signal is present.

In the transistor, there will be a base current I_B of 2 mA flowing, and this current will also flow through the resistor R_B. In the absence of signal, there will be no other current flowing in the base lead.

In the valve, there will be a potential difference of 2 volts between the control grid and the cathode, the grid being negative with respect to the cathode. Under these circumstances, the current flowing through the valve (I_a) is found from the characteristics to be about 14 mA. This current also flows through R_B.

If the bias resistor R_B in the valve cathode lead is 142 ohms, then the 14 mA flowing through it will give a potential difference of 1·988 volts between cathode and earth. The cathode is thus approximately 2 volts positive with respect to earth. The grid is connected to earth through the grid leak resistor R_g (see below), and in the absence of any signal voltage there is negligible current flowing in this resistor and the grid is approximately at earth potential. The cathode is thus 2 volts positive with respect to the grid, i.e. the grid is 2 volts negative with respect to the cathode. $V_g = -2$ volts, the situation we require.

The required bias voltage is obtained directly from the h.t. supply in the valve amplifier by the use of a cathode bias resistor $R_B = 142$ ohms. In practice, a nominal 150-ohm resistor would be used in the circuit.

In the transistor amplifier, we require to choose the value of R_B so that the current flowing in it (I_B) is 2 mA.

One end of this resistor is at h.t. potential; the other end is connected to the base. Because the emitter-base junction is biased in the forward direction, the potential drop across it is small and, for this calculation, the base and emitter can both be regarded as at earth potential. Thus one end of R_B is at h.t. potential and the other at nearly earth potential. The value of R_B can be found with sufficient accuracy by dividing the h.t. voltage by the bias current required:

$$R_B = \frac{25 \text{ volts}}{2 \text{ mA}} = 12\,500 \text{ ohms.}$$

In the valve circuit above, there is a resistor R_g connected between grid and earth. This resistor is called the grid leak, and its purpose is to ensure that there is a direct path for current to flow from grid to earth. Such a path must be provided in triode, pentode, etc., valves because some of the electrons crossing the valve from cathode to anode will inevitably strike the wire of the control grid. If no grid leak were provided, there would be a gradual accumulation of negative charge on the grid and the working point of the valve would be altered.

The value of the grid leak resistor is not critical and is usually about a hundred kilohms so that there is no possibility of it short-circuiting, or appreciably affecting, the source delivering the signal to the amplifier. The leakage current flowing in R_g is extremely small and the d.c. potential difference across the grid leak is usually regarded as negligible. In particular, it is not considered to make any contribution to the bias, and the grid is taken to be at earth potential when no signal is present.

Modified Amplifier Circuits

Fig. 69 shows the circuit of a valve amplifier which has an extra component—a capacitor C_B, in parallel with the bias resistor R_B, between cathode and earth. The function of this extra component is to prevent the occurrence of something called negative feedback, which is described below.

R_B was chosen so that the designed steady current flowing through the valve would give a potential difference across R_B just sufficient to provide the correct bias required to make this designed current flow. In the example above, a steady current of 14 mA flowing in the valve gave 2 volts potential difference across the 150-ohm cathode bias resistor. The grid is 2 volts negative with respect to the cathode, and this is the steady working point of the amplifier.

When a sinusoidal signal is applied to the circuit, one half-cycle will take the grid potential closer to that of the cathode and the other half-cycle will take the grid farther away from the cathode potential. These fluctuations in the grid–cathode potential difference will cause the current in the valve to rise and fall so that an alternating current is superimposed on the original steady current

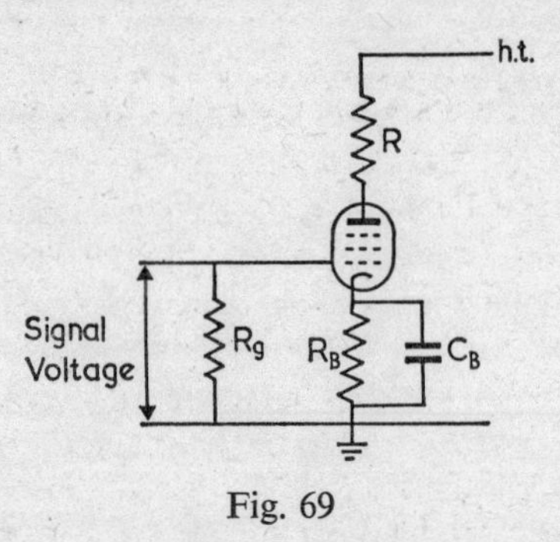

Fig. 69

flowing in the load. The alternating component of the voltage across the load is the amplifier output.

When the signal makes the grid potential come closer to the cathode, the current through the valve increases. But this increase in current increases the potential difference across R_B and, in particular, the cathode potential relative to earth alters. When the signal makes the grid come closer to the cathode in potential, the consequent change in potential across R_B tends to take the cathode potential away from the grid. The net change in potential between grid and cathode is less than would have been expected because the change in potential across R_B opposes the change produced by the signal voltage.

The output voltage of the amplifier is due to the alternating component of the valve current passing in

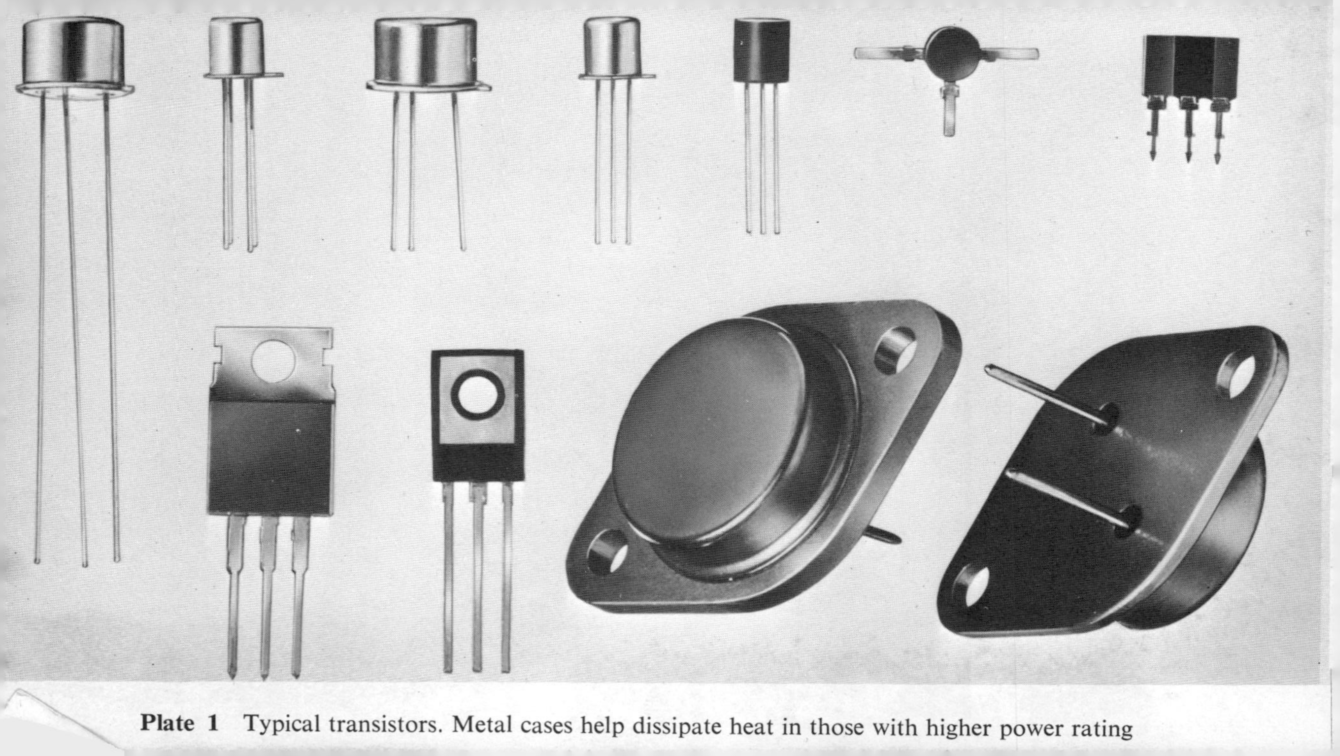

Plate 1 Typical transistors. Metal cases help dissipate heat in those with higher power rating

Plate 2 Integrated circuit with protective capsule and connecting pins (Similar scale to Plate 1)

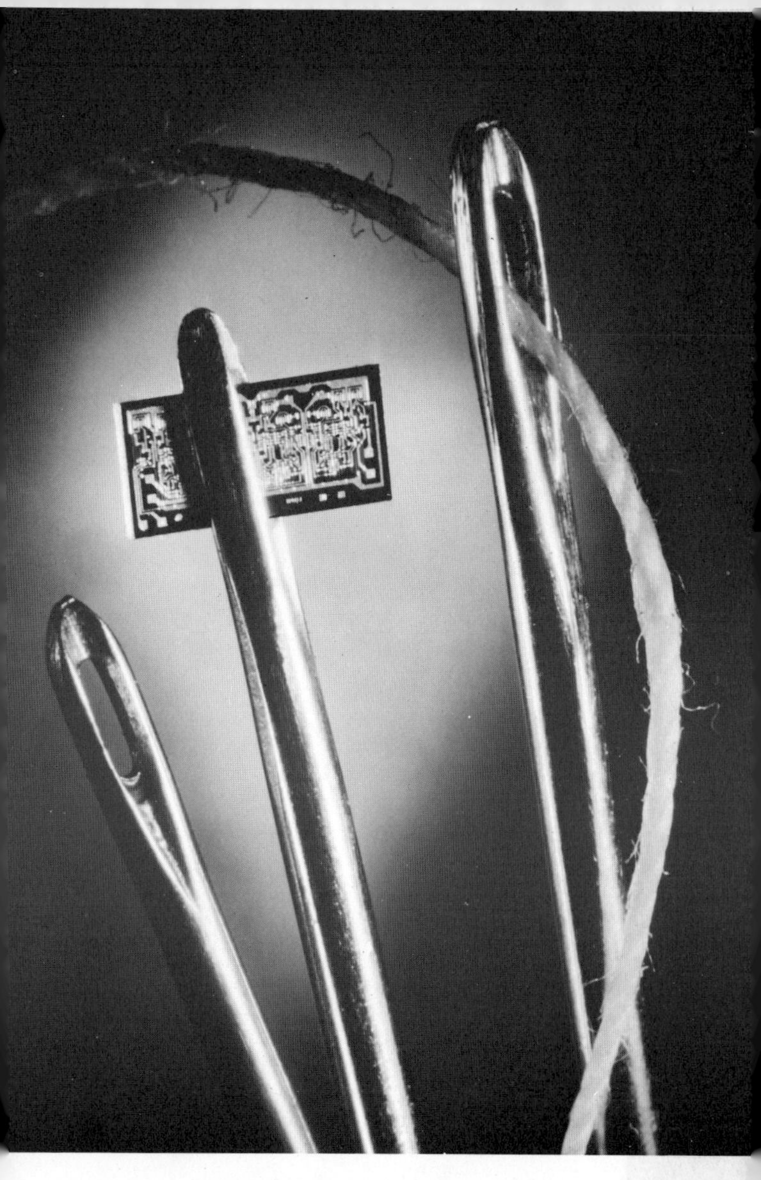

Plate 3 Magnified picture of integrated circuit formed on single silicon chip

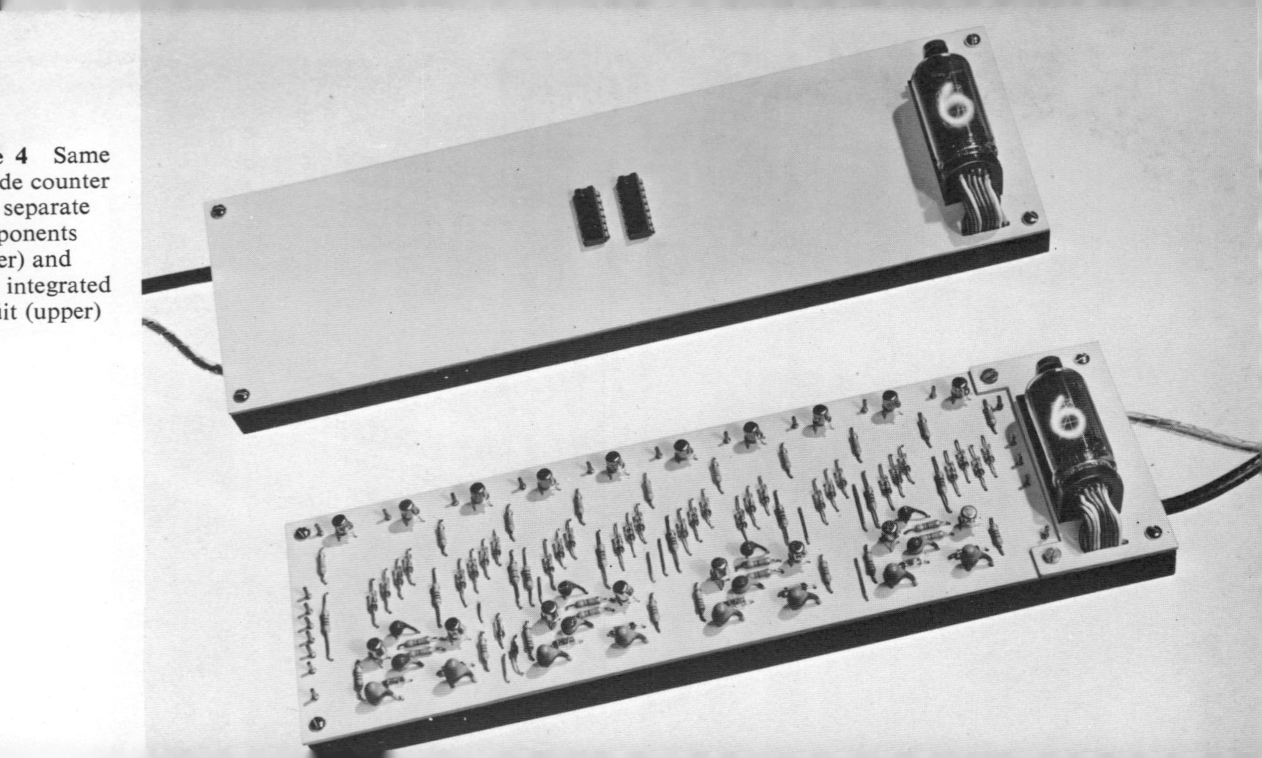

Plate 4 Same decade counter with separate components (lower) and with integrated circuit (upper)

series through the load R. In the simplest amplifier, this alternating current arises because the input signal to the amplifier is an alternating voltage which is applied directly between the grid and cathode of the valve. But when R_B is introduced into the cathode lead, the alternating component of the valve current flowing through this impedance produces an alternating voltage between cathode and earth which opposes the input signal. The effective alternating voltage, i.e. the actual alternating voltage between grid and cathode, is the amplifier input signal minus the alternating voltage produced across the impedance between cathode and earth (R_B).

The a.c. voltage across R_B opposes the signal and reduces the effective input. The output of the amplifier is also reduced because there will not be so great a fluctuation in valve current as would have been expected. Thus the amplifier output for a given signal applied to its input terminals is reduced, i.e. there is a reduction in amplifier gain.

The a.c. voltage generated in R_B opposes the input to the amplifier and reduces the effective gain. Such a process is called negative feedback.

R_B was introduced into the circuit to provide certain d.c. conditions, i.e. the correct steady bias voltage. But it is now found to produce unwanted a.c. effects since it also introduces an a.c. impedance between cathode and earth, across which a negative feedback voltage appears when a signal is applied to the amplifier. However, if a capacitor C_B is placed in parallel with R_B, the d.c. resistance between cathode and earth remains unaltered, but the a.c. impedance is reduced.

The a.c. impedance of a capacitor is $\frac{1}{2\pi fC}$ ohms, where f is the frequency of the alternating voltage in Hz and C is the capacitance in farads.

If $f = 1$ kHz and $C = 100$ microfarads, then the impedance of the condenser is

$$\frac{1}{2\pi \times 1000 \times 100 \times 10^{-6}} = \frac{10}{2\pi}$$

which is less than 2 ohms.

If this capacitor is placed in parallel with R_B, then the impedance between the cathode and earth is reduced from the original 150 ohms, due to R_B, to something less than 2 ohms and the negative feedback becomes negligible.

Note that, to be equally effective when the signal frequency is 10 Hz instead of 1000 Hz, the capacitance would have to be increased by a factor of 100.

The implication of the above discussion is that negative feedback is to be avoided because the only important feature of an amplifier is its gain. But should gain be important in a particular application, it is easy to add one or more stages of amplification to achieve the figure required. At least of equal importance in many cases is the bandwidth or the stability of the amplifier.

The bandwidth of an amplifier is the range of frequencies which can be satisfactorily amplified. Each circuit has upper and lower limits of frequency beyond which the gain falls off sharply. With any particular valve or transistor, the reduction in gain by the use of a circuit giving negative feedback is matched by a corresponding increase in bandwidth. In some circuits, e.g. a high fidelity audio-frequency amplifier, a large bandwidth is particularly desirable and the condenser C_B will be omitted so that negative feedback occurs. Furthermore, the presence of negative feedback tends to remove any possibility of unwanted oscillations in the amplifier due to stray signals leaking into the amplifier input from the high power output side (see oscillators, page 152).

There is a special sort of instability against which R_B is effective, even when it is used with C_B in parallel to eliminate negative feedback at signal frequencies. This is the drift of the working point of the valve from the selected value because of long-term variation in the supply voltage, change in cathode emission current with age and effects due to changes in ambient or component temperatures. Such drift is so slow that the presence of C_B does not remove the negative feedback effect because the impedance of the capacitor for such an extremely low-frequency voltage is very high and R_B is no longer short-circuited for a.c.

If any of the above effects tend to make the steady valve current increase, the potential difference across R_B will increase and the cathode will become more positive with respect to the grid, i.e. the grid will become more negative with respect to the cathode and the valve current will therefore be reduced.

Stabilisation of the working point in this manner is especially important in transistor circuits, which are much more susceptible than valves to temperature changes. For this reason, an extra resistor R_E is frequently introduced into the emitter lead with a capacitor C_E in parallel (Fig. 70). The standing d.c. potential difference across R_E, due to the steady current flowing in the absence of signal, represents a small reduction in the effective h.t. voltage and this should be taken into account in designing the circuit.

Two-stage Amplifiers

The circuits shown in Fig. 70 below are two-stage amplifiers. In the transistor amplifier, the emitters are marked *E* to avoid using the arrow, which also distinguishes *p–n–p* from *n–p–n*, because either type of transistor may be used in the circuit as long as the correct polarity is used for the h.t. supply. The components labelled are the only ones whose function has not already been described.

In the valve circuit are shown the extra connections

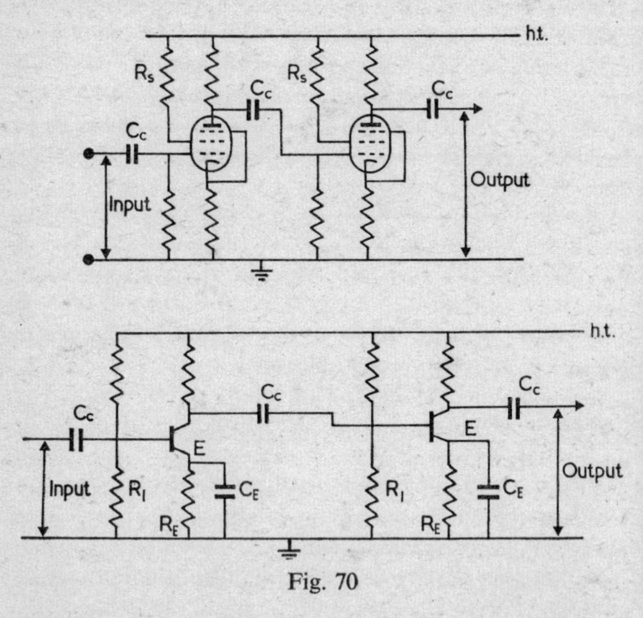

Fig. 70

needed when pentodes are used. The suppressor grid—that nearest the anode—is normally connected directly to the cathode. The screen grid may be connected to the h.t. directly or through a resistor R_s, as shown. The end of R_s attached to the screen is sometimes also connected to earth through a condenser so that the potential of this grid cannot fluctuate when a.c. currents flow in the circuit.

The capacitors C_C in both circuits are called coupling capacitors. Their purpose is to allow the a.c. signal to pass between two points, but to prevent the flow of d.c. between these points. Thus a coupling capacitor is used to take the signal from the anode or collector of the first stage to the grid or base of the second stage. But the very different d.c. potentials present at the output of the first stage and at the input of the second stage do not interact because the capacitor is an open circuit for d.c.

The coupling capacitors at the input of the first stage and the output of the second stage are not always necessary. Their function is to isolate the d.c. potential at the grid or base of the first stage and at the anode or collector of the second stage from any d.c. circuits or potentials which are to be connected to the input or output of the amplifier.

The resistor R_1 in the transistor amplifier helps in the stabilisation of the working point. Together with the resistor connected directly from base to h.t. it constitutes a potentiometer, maintaining the correct potential difference between base and earth.

Conclusion

In this chapter we have been concerned with the design of the fundamental electronic amplifier. The design of this amplifier can be approached in other ways: perhaps using a similar graphical method, but with different characteristics, or possibly a purely mathematical approach with the characteristic curves represented by equations.

In Chapter 8 we shall see how this design can be modified to produce different sorts of amplifier, and also how amplifiers can be modified to carry out functions other than amplification.

CHAPTER 7

Switches

The perfect switch is a piece of equipment which can be made to have either zero resistance or infinite resistance between two of its terminals at the will of the operator.

If a thick metal bar has a break in it which is filled when an operator inserts a key, then this constitutes a simple mechanically operated switch. Developments of such a

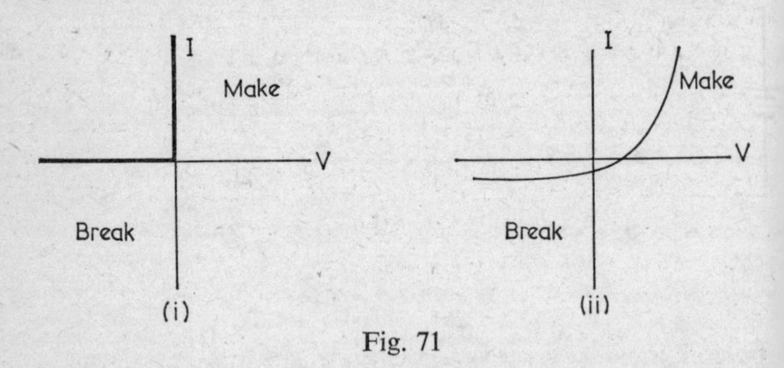

Fig. 71

switch are commonly used for making and breaking the supply of power to lights, motors and electrical equipment generally. Such switches come very close to the perfect switch defined above, but they are all used in applications where the switching action never has to be performed faster than a few times per second at most.

In this chapter we shall discuss electronic switches using diodes, amplifying valves or transistors.

Such switches have characteristics of the general form

of Fig. 71 (ii) rather than the perfect characteristic of (i). But, in compensation for the disadvantages of the characteristic (e.g. power is consumed in the switch), electronic switches can be operated at extremely high speeds and can thus be used in applications, like computer logic, which are discussed in Chapter 9.

Diode Switch

Consider first the circuit of Fig. 72, where a resistor R is placed in series with a mechanical switch. The ends of the circuit, B and C, can be maintained at various fixed

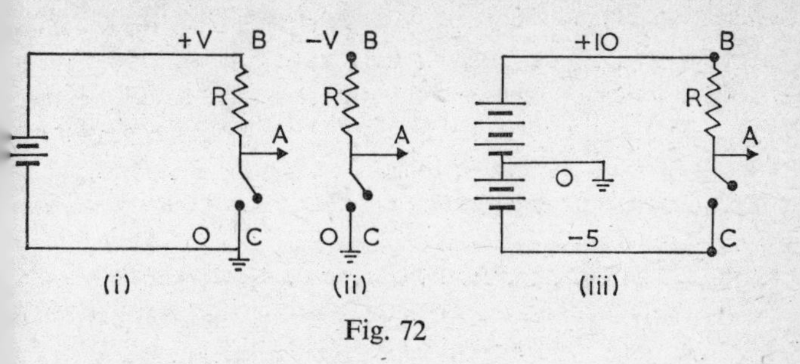

Fig. 72

potentials, as shown, by the use of battery or other power supplies.

In (i), when the switch is broken, no current flows in the circuit and there will thus be no potential difference between the ends of the resistor. The point A will be at the same potential as the point B, i.e. $+V$ volts.

When the switch is made, the circuit is completed and current will flow. But the switch has zero resistance, so there will now be no potential difference between the points A and C, i.e. A will be at earth potential.

When the switch is open, A will be at $+V$ volts, and

when the switch is closed the potential of A will change to 0 volts.

In (ii), when the switch is open, A will be at $-V$ volts, and when the switch is closed the potential of A will change to 0 volts.

In (iii), when the switch is open, A will be at $+10$ volts, and when the switch is closed the potential of A will change to -5 volts.

Let us now consider similar situations with the mechanical switch replaced by a diode.

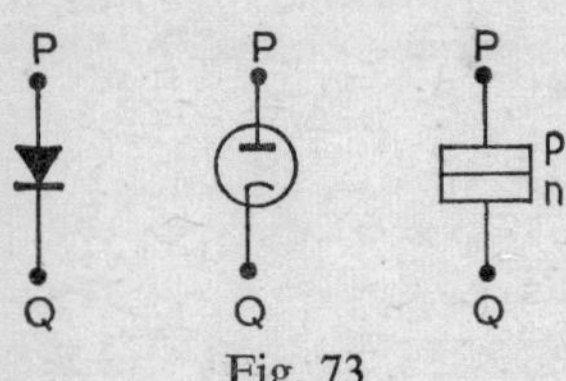

Fig. 73

The diode may be a solid state device or a thermionic valve, but it will be represented by the conventional rectifier symbol, shown in Fig. 73, with the equivalent junction and valve diodes. In each case, the easy flow or forward direction is with P positive with respect to Q.

We shall, for the moment, assume that when the diode conducts there is a negligible potential difference across it, and that in the reverse direction zero current flows. The diode is assumed to be a perfect switch.

The principles of operation remain the same as with the mechanical switch, i.e. the potential of A equals that of C when the diode conducts and the potential of A equals that of B when the diode does not conduct.

There is, however, one important extra principle. The diode will always conduct when A is made positive with respect to C; it will never conduct unless this is so.

Once the potentials at A and C have been established, the state of the diode—conducting or non-conducting—will have been determined and there is no way for an operator to control it separately.

In Fig. 74 the mechanical switch has been replaced by a diode. In (i) and (iii) the diode will conduct because A is made positive with respect to C. The potential at A will be (i) 0, (ii) $-V$ and (iii) -5. The potential of A in (i) and (iii) will be slightly more positive than the values

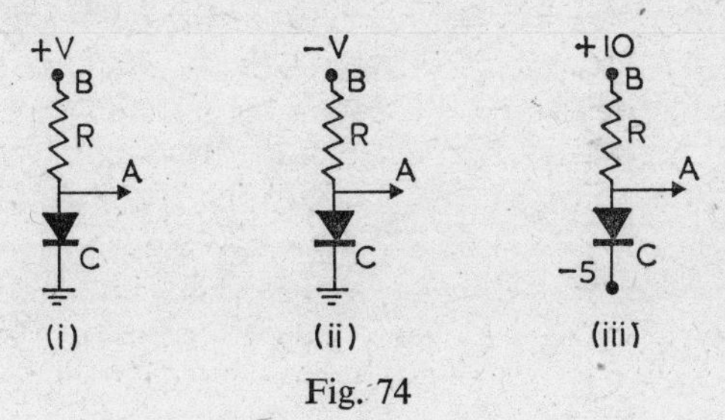

Fig. 74

quoted because of the potential difference which exists across a real diode when it conducts.

Ignoring the potential difference across the conducting

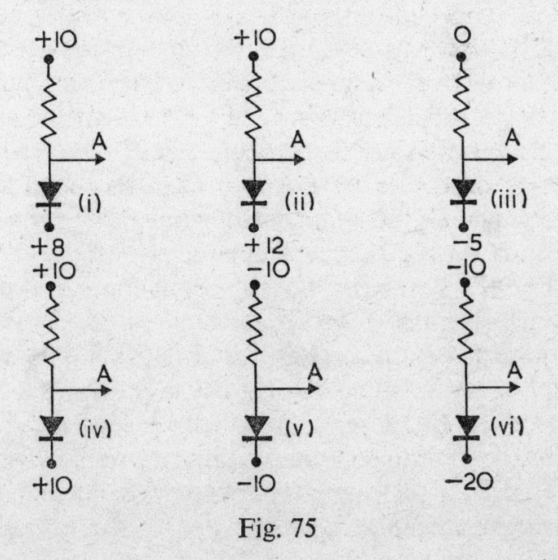

Fig. 75

diode, note on Fig. 75 the potential at *A* in all the cases. The answers are given on page 141.

Let us now consider what happens to the potential at *A* when the potential at one or both ends of the circuit changes with time in a certain manner. We shall be particularly concerned with voltage inputs which are rectangular pulses or sinusoidal waves. Once again, we shall assume that the diode is an ideal one with no significant potential drop across it when it conducts.

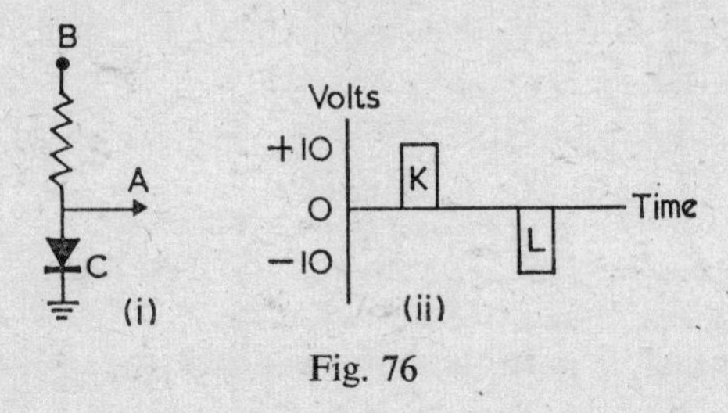

Fig. 76

In Fig. 76, *C* is held at earth potential, and the potential of *B* relative to earth is shown in (ii). Thus *B* is normally at earth potential, but positive going voltage pulses like *K* or negative going pulses like *L* can be applied to it.

When no pulse is present, *B* is at earth potential, and so are *C* and *A*. When a positive pulse (*K*) arrives, the diode conducts and remains conducting for the duration of the pulse. While the diode is conducting, *A* will be at the same potential as *C*, i.e. earth potential. As soon as the pulse has ended, *B* returns to earth potential, the diode stops conducting and the original state of affairs is restored, with *A*, *B* and *C* all at zero volts.

Thus, if a positive voltage pulse is applied to *B*, there is no change in potential at *A* and no pulse is available on an output lead connected to *A*.

When a negative pulse (*L*) is applied to *B*, the diode will at no time conduct. The potential at *A* will at all times be the same as that at *B*, and the negative pulse is available on the output line.

If the potential at *B* is regarded as the input and that at *A* as the output of the circuit, then the negative going input voltages are transmitted unchanged, but positive going voltages are not transmitted.

If the positions of the resistor and the diode are reversed, as in Fig. 77, then positive going voltage inputs at *B* will make the diode conduct and the point *A* will follow *B* in potential. For negative input voltages, the diode will not conduct and *A* will remain at the same potential as *C*, i.e. earth potential. This circuit transmits positive voltage signals, but not negative ones.

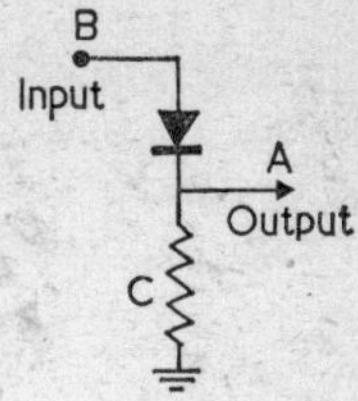

B
Input
A
Output
C

Fig. 77

In Fig. 78 we have returned to the original arrangement of Fig. 76, but with *C* biased to —2 volts instead of at earth potential. In this case, the diode will always conduct and *A* will thus be at —2 volts, as long as the point *B* is at any potential more positive than —2. If, however, an input is supplied to B which makes it more negative than —2 volts, the diode will stop conducting and the point *A* will follow *B* in potential. Any input voltage, or portion of it, below —2 volts will be transmitted.

Thus, if *B* is normally at earth potential and a negative going input pulse of 10 volts amplitude is then applied to it, a negative output pulse of 8 volts amplitude will be obtained.

Answers to Fig. 75:
(i) +8 (ii) +10 (iii) —5 (iv) +10 (v) —10 (vi) —20.

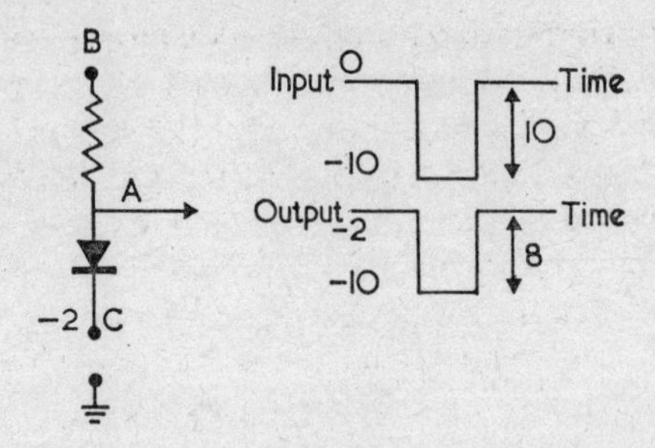

Fig. 78

In Fig. 79 two diodes are used in the circuit, which is an important one in computer logic and is discussed in Chapter 9.

Suppose that the two input lines are normally at earth potential, but that positive pulses may appear at B_1 and B_2. In the absence of any pulses, both diodes will conduct and the output A will be at earth potential—assuming the diode potential drop to be zero. If a positive pulse appears on one of the input lines, the diode in that line will stop conducting for the duration of the pulse, the height of which is arranged to be of much greater amplitude than 6 volts. But the diode in the input line which has no pulse will still be conducting, so the potential of A will remain held at earth potential. There will be no change in potential at A, and therefore no output, if an input pulse appears on one line only.

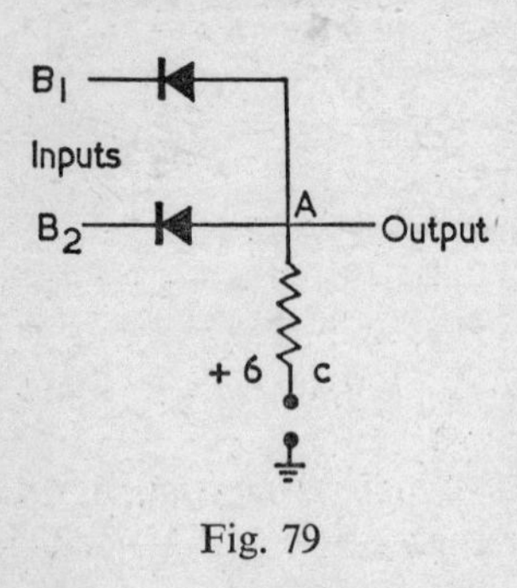

Fig. 79

If, however, a positive pulse appears simultaneously

on both the input lines, both diodes will stop conducting, and the point A will rise from zero to $+6$ volts and remain there until the input pulses end. The diodes will then conduct again and A will return to earth potential. There will be a positive pulse delivered from the output only if a positive pulse appears simultaneously on both input lines. This circuit is called an AND-gate.

Real Diodes

In the above circuits, the diode has been considered to be a switch which has no current flowing in it when off and no voltage drop across it when on. Such simplification is normally quite justified in understanding the main working of switching and logic circuits. But when considering the details of practical circuits it will be necessary, for instance, to know how big the input pulses must be to make a switch function, and how big will be the consequent output pulses; how big must the biasing voltage be, and what variation is possible before the switch no longer functions correctly; and what will be the effect of one circuit upon its neighbour when the output of one is fed to the input of another in a complicated logic process.

To establish such points of detail it must first be specified whether thermionic or solid state diodes are to be employed. With thermionic diodes, there will be no reverse current, but the forward voltage drop when conducting may be quite high. It is much more likely that solid state diodes will be used in most switching applications because their small size, simple construction and ease of connection allow a very large number to be packed cheaply into a small space. Whatever diode is used, the necessary details of operation can usually be established by manipulating load line and characteristics in the manner described in Chapter 5.

Transistor and Pentode Switches

Greater flexibility in switching can be achieved at the expense of slightly greater circuit complexity if transistors

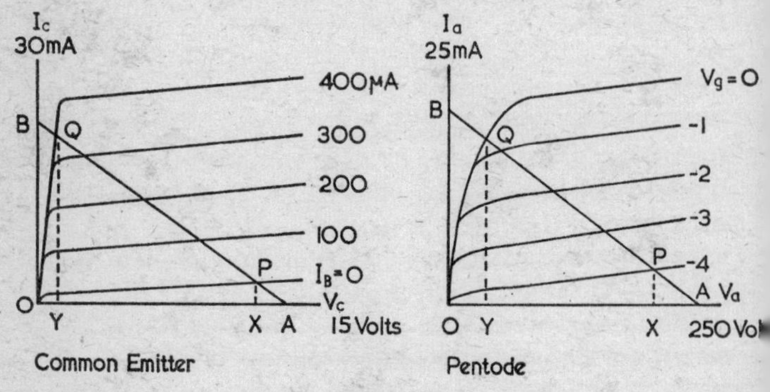

Fig. 80

or pentodes are used instead of diodes. The switch can now be opened and closed by applying a suitable signal to the base or control grid, and the amplifying action allows the height of pulses to be maintained, in spite of losses of voltage in the resistors in logic circuits.

On the characteristics shown in Fig. 80, the load line has been drawn for a resistor in series with the valve or transistor in a switching application. Notice that, in contrast to the amplifier, the load resistor in the anode or collector circuit is now chosen so that the load line intersects the steeply rising part of the characteristics.

The working point is switched between P and Q by providing an input which changes I_B or V_g in the appropriate manner indicated by the characteristic. When the working point is switched from P to Q, the collector or anode potential changes from X to Y, and the output pulse is in general an amplified, inverted version of the input pulse.

We shall see in Chapter 9 that switches of this sort, particularly transistor switches, are used for AND-gates and OR-gates in place of diodes, and that they are very frequently met in pairs and bigger groups as pulse generators and counters.

Switching Speed

In discussing the action of various switches, especially those to be used in logic circuits, we have considered rectangular pulses as input and output. If a pulse is assumed to be rectangular, then it is implied that the potential at some point in the circuit, e.g. the input or output of the gate, changes from one value to another in zero time. This can never be the case in practice and, in digital computers especially, the switching speed is an important design consideration.

If the potential at any point in the circuit is to change, then electric charge must flow into or away from it. There will be resistance (R) associated with the path the charge flows through and capacity (C) associated with the point to which it is flowing. The time taken for a change of potential to take place depends on the product CR, and this must be kept as small as possible by careful choice of components and by attention to the circuit layout to avoid unwanted stray capacity. Certain other delays arise from physical processes within the semiconductors normally used in preference to thermionic valves in high speed computing circuits. The time taken for charge carriers to diffuse through the material may well be significant, and manufacturing processes have been improved to reduce such delays to very low values so that fast diodes and transistors built into well designed circuits will give switching times of the order of a nanosecond (10^{-9} s).

CHAPTER 8

Applications of Amplification

In the description of amplification in Chapter 6, the underlying assumption was that an amplifier was required to accept a signal of very low power—from, say, an aerial, a gramophone pick-up or a photo-cell—and to increase the power level sufficiently to operate a telephone receiver, a loudspeaker or a mechanical relay.

Practical amplifiers often have several stages, the final stage of amplification having to provide the power to drive the loudspeaker or other output device. Often, therefore, this final stage will have a bigger valve or transistor than the rest of the amplifier and will be referred to as a power amplification stage. If the amplifier uses thermionic valves, the last one will require a large signal voltage applied to its grid to generate the necessary large power output. The function of the earlier stages, then, is to amplify the original signal input voltage until this requirement is satisfied; for this reason, they are sometimes referred to as voltage amplifiers.

But the most important classification of amplifiers, which we shall consider below, is in terms of frequency and bandwidth. In particular, we shall consider the tuned amplifier, which only responds to a comparatively narrow band of frequencies.

The phenomenon of negative feedback has been met in Chapter 6, where it was found to reduce the gain of an amplifier and increase the bandwidth. In this chapter we shall see the application of positive feedback to convert an amplifier into an oscillator, which is a generator of alternating electrical power at a frequency determined by the components of the circuit and thus, among other things, can be the basis of a radio transmitter.

Bandwidth; Tuned Amplifier

On page 134 the two-stage valve and transistor amplifier with resistive load were described. If the voltage gain of such an amplifier is measured for various different signal frequencies, then the frequency response has the general form shown in Fig. 81.

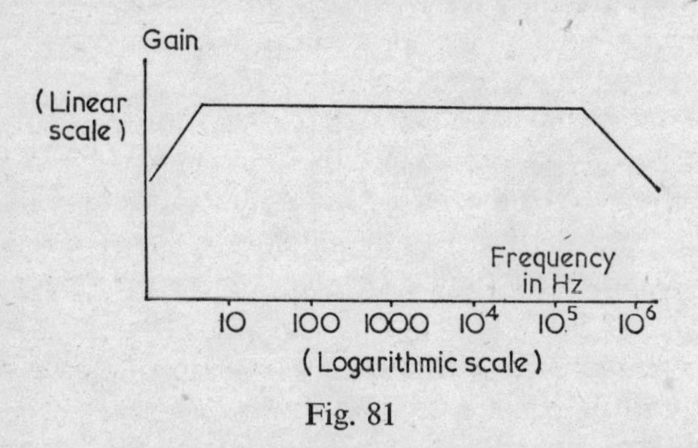

Fig. 81

The gain is constant over a wide range of frequencies, but falls off at very low and very high frequencies. These falls in gain are caused by capacitance in the circuit. The low-frequency fall is due to an actual component, the coupling capacitor C_C, whereas the high-frequency fall is due not to a real component but to an equivalent capacitance C_E, which effectively appears across the input terminals (see Fig. 82).

In Fig. 82 are shown the significant parts of the circuit of a two-stage resistance amplifier. R is the load of the first stage, R_1 is the resistor across the input to the second stage and C_C is the coupling capacitor intended to take the signal from the first to the second stage without loss. C_E is not a component, but represents certain physical phenomena which occur in the valve or the transistor and have the same effect upon the performance of the circuit as C_E, which has a small capacitance, possibly a few picofarads for a valve and a few hundred picofarads for a transistor.

At very low frequencies, the impedance of C_C becomes significant (see page 131) and some of the output voltage

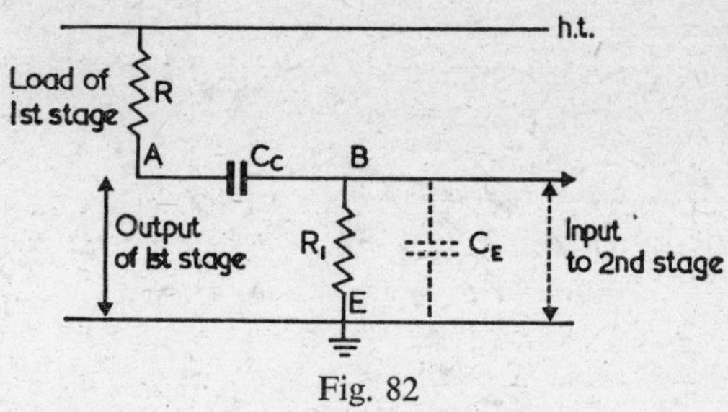

Fig. 82

of the first stage, which is developed between A and earth, is dropped between A and B. The input to the second stage is thus less than it should be and the total gain of the combined stages falls.

At high frequencies, the impedance of C_E falls low enough to be comparable with the resistance of the load R of the first stage. Because the h.t. line and earth are the same point as far as a.c. is concerned, C_E is effectively in parallel with the load of the first stage. When the impedance of C_E becomes comparable with R, the load of the first stage is effectively reduced and the gain falls.

Even with the fall at high and low frequencies, the amplifier with resistive loads will amplify a wide band of frequencies, and with negative feedback the bandwidth can be further increased at the cost of lowering the gain.

Where gain is important at the very lowest frequencies e.g. a very small fraction of a Hz, then the circuit may be so designed that no coupling capacitor is needed to keep the steady potentials on the two stages separate. Such amplifiers are called direct-coupled amplifiers (often abbreviated to D.C. amplifiers) and are used in special applications associated with process control, analogue computation, etc.

The gain of an amplifier depends very closely upon the

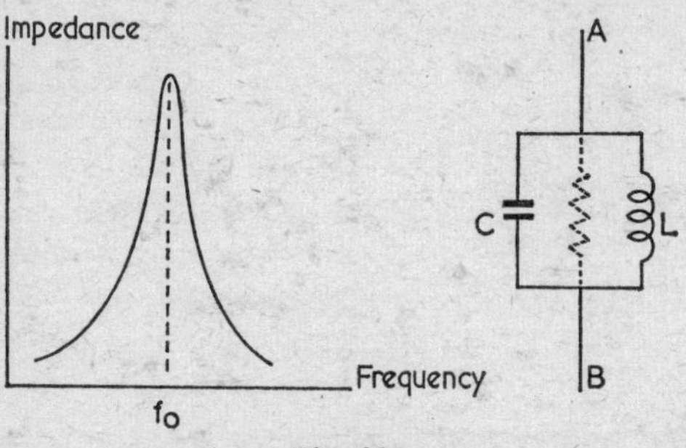

Fig. 83

impedance of its load and, if this load is a resistor, the impedance will not vary with frequency. The amplifiers so far described have resistive loads and their gain is independent of the signal frequency over a large band-width, only falling off at low and high frequencies due to secondary circuit effects. There is, however, a requirement for an amplifier which is selective, i.e. which only amplifies a certain narrow band of frequencies. To achieve this the simple resistive load is replaced by a parallel tuned circuit, whose impedance depends strongly on frequency.

A parallel tuned circuit consists of a capacitor *C* and an inductor *L* in parallel, as illustrated in Fig. 83. The dotted resistor shown is not a component in the circuit, but represents the effect of the losses. The impedance between A and B then has the form shown, rising to a peak at f_o, the resonant frequency, where $f_o = \frac{1}{2\pi\sqrt{LC}}$ (L in henries, C in farads, f_o in Hz).

In fact, the impedance should rise to infinity at resonance, but in any real tuned circuit there are some

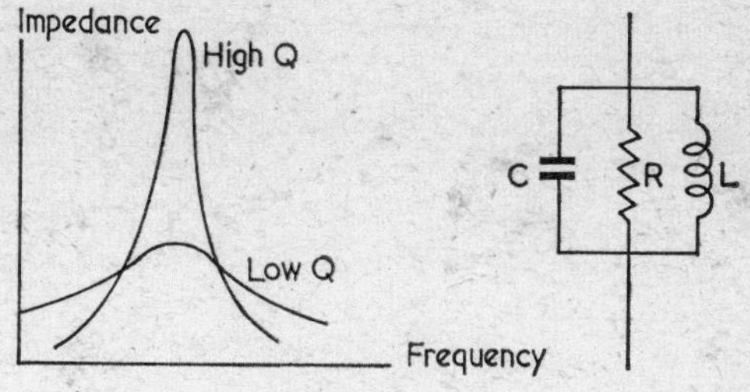

Fig. 84

losses when the alternating current flows through the inevitable resistance of the inductor, connecting wires, etc. If these losses are high, the resonance curve is flatter and the circuit is said to have a low Q factor (Q for quality).

The responses of high and low Q circuits are shown in Fig. 84. Occasionally, the resistive losses in a circuit are deliberately increased by placing a resistor R in parallel, as shown, so that the circuit will become less sharply tuned.

If the load resistor of an amplifier is replaced by a parallel tuned circuit, then it becomes selective and amplifies very strongly signals of the resonant frequency and a narrow band on either side. Other signals are not amplified, and such amplifiers will be used in radio receivers to select the signal from one particular transmitter when signals at many other frequencies are present at the aerial.

In Fig. 85 the relevant a.c. portions of a two-stage tuned amplifier are shown with parallel tuned circuits as loads. The connection between the two stages could have been made through a coupling capacitor as before, but in the case shown the output of the first stage is transformer-coupled to the input of the second stage.

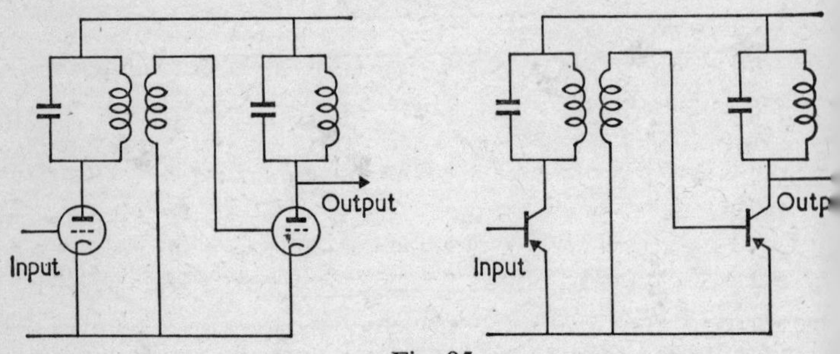

Fig. 85

A multi-stage amplifier of this type, with all the loads tuned to the same frequency, will be very sharply selective. Occasionally, successive stages will be tuned to very slightly different frequencies so that the overall response of the amplifier has a wider bandwidth centred on the required frequency.

Positive Feedback; Oscillation

In the tuned amplifiers described in the last section, there was transformer coupling between stages, i.e. a small coil connected to the input of the second stage was placed near the coil in the load circuit of the first stage so that power transfer took place by mutual inductance between the two coils.

Suppose we now take a single tuned amplifier stage and again place a small coil near the coil in the load circuit. When a signal is applied to the input terminals, an amplified version will appear in the output circuit and a voltage will be induced in the small coil.

Let us now take the small coil and connect it to the input terminals of the amplifier, at the same time removing the original input signal. If the voltage which had been induced in the small coil was of the same amplitude and in the same phase as the original signal, no difference would be apparent to the amplifier and it would continue to produce alternating power in the output circuit. With no input signal, the 'amplifier' is now continuously generating alternating power at the frequency of the tuned circuit. It is an oscillator.

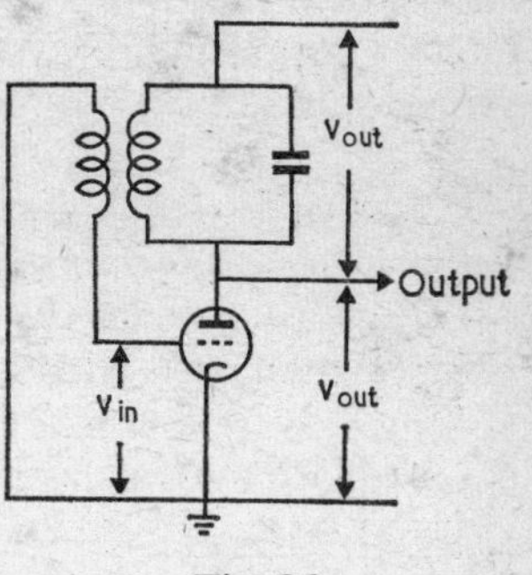

Fig. 86

The h.t. supply is the source of the alternating power now available in the load circuit. The great importance of the invention of the oscillator was that it provided a controlled source of radio-frequency oscillations at a particular frequency. It became the basis of the modern sophisticated radio systems that have superseded the original wireless communication by on–off keying of a spark discharge.

In the oscillator illustrated in Fig. 86, the significant a.c. circuits are shown, and the active element might equally well be a transistor instead of a valve.

The output voltage of the oscillator is V_{out}, which is

the voltage across the load and, in particular, across the ends of the coil in the tuned circuit.

By transformer action, a fraction of this alternating voltage is induced in the small coil, and this is applied to the input terminals. This voltage is v_{in} and hence the gain produced by the active element is $G = \frac{v_{out}}{v_{in}}$.

For oscillation to occur, it appears that the coupling between the two coils must be such that the feedback voltage is exactly $1/G$ of the output voltage and that the phase of this input voltage must also be correct.

In reality, the requirements are not as stringent as this because the circuit is to a certain extent self-adjusting.

Let us first consider how oscillations are set up in a practical oscillator circuit where the feedback coil is already connected to the input of the valve or transistor. The sleight of hand by which an external signal was replaced by the feedback voltage will not be invoked, as it was in the explanation given a few paragraphs earlier.

First, it will be arranged that the coupling between the two coils is closer than suggested above so that a fraction greater than $1/G$ of the output voltage is fed back. Suppose that the gain is 100 and that a voltage equal to 2 per cent of the output is fed back.

When the oscillator is first switched on, the electrical shock will cause oscillations of minute amplitude to occur in the tuned circuit at its resonant frequency. In the same way, the mechanical shock from a slammed door will cause a number of pendulums hanging on the wall to oscillate slightly, each at its own resonant frequency.

The minute oscillating current in the tuned load will mean that there is a very small alternating voltage across it, and 2 per cent of this voltage will be fed back and applied to the input terminals. Since the gain is 100 a voltage, twice the amplitude of the original shocked oscillation will now appear in the tuned circuit.

The new input will be 2 per cent of this larger voltage and there will be a corresponding increase in the amplitude of the oscillations in the tuned circuit. Thus, when the oscillator is first switched on, there will be a rapid build-up of oscillations, provided that the original percentage feedback exceeds a certain value, which will be small in high gain circuits.

The ultimate amplitude of the oscillation is limited by the fact that the valve or transistor will be taken into its non-linear region when the voltage becomes large. In the non-linear region, the gain becomes lower, and hence the greater the excursion of the circuit into the non-linear region, the smaller the mean gain becomes. The amplitude of oscillation builds up until the excursion into the non-linear region is just sufficient to reduce the effective gain to a value at which the percentage feedback will maintain this amplitude at a constant value, i.e. in the simple case above the mean gain will be reduced to 50.

There is a fundamental requirement that the feedback coil be connected the right way round to the input of the valve or transistor. If the connections are reversed, the phase will be changed by 180 degrees and oscillation will not be possible. Apart from this, it is not important that the phase relation should be made precisely correct when the circuit is constructed. Provided that there is sufficient feedback, the voltage applied to the input terminals will have a component in the correct phase large enough to produce oscillation, even if its own phase is not exactly correct.

In some oscillators the feedback coil has a capacitor placed in parallel with it, so that there is a tuned circuit in both the input and the output side of the valve or transistor. In general, both the circuits will be tuned to the same resonant frequency, but in practice there will usually be some slight difference. The actual frequency of oscillation is determined by the circuit which has the higher Q.

In broadcast transmitters it is important that the frequency should not vary with time, and in such applications one of the tuned circuits might be made with great care to have a high Q. In operation, this circuit could be maintained under constant temperature conditions in an oven so that good frequency stability was achieved. A suitably cut slice of quartz crystal with metal plating on two opposite faces can be made to behave in many respects like a tuned circuit when an alternating voltage is applied between the two faces. The effective tuned circuit has an extremely high Q, and crystal-controlled oscillators, with special feedback arrangements replacing the transformer coupling considered above, are commonly used when high stability is required.

Efficiency of Amplifiers and Oscillators

The efficiency of an amplifier or an oscillator may be defined as the ratio of the alternating power output to the total power input. Several items, such as valve heater power, might be included in the total input power, but the principal component is the d.c. power drawn from the h.t. supply.

Efficiency is an important practical consideration in high power circuits, where a few per cent difference in efficiency can mean a large absolute amount of power. This is expensive and, perhaps even more costly, wasted power must be dissipated from the circuit as heat. In battery operated apparatus, efficiency is likely to be closely considered even at low power levels.

We have already given some consideration to amplifier efficiency (page 125), where we saw that the steady working point in the absence of signal would be chosen at as low a current as was practicable, since this would represent the mean current drawn from the h.t. supply. The circumstances considered were such that there was a sinusoidal current flowing in the valve or transistor, with a sinusoidal voltage at the input and output terminals. This is sometimes called Class A working, when current flows in the active element during the whole 360 degrees of the input or output voltage cycle.

By lowering the working point, the mean current taken from the h.t. supply is reduced and the efficiency increased. But if the process is continued, one half-cycle of the input voltage will extend beyond cut-off, current will flow for less than 360 degrees and unacceptable distortion will generally occur.

There are, however, certain circumstances under which an 'angle of flow' of less than 360 degrees can be used, with the advantage of higher efficiency but without the disadvantage of distortion.

The parallel *LC* circuit, which is the load of a tuned amplifier, is a resonant system and a continuous sinusoidal voltage at the resonant frequency will appear across it, even if current only flows into it from the valve or transistor in a short pulse once every cycle. In the same way, it is not necessary to run behind a child on a swing pushing, or pulling, the whole time. A sharp push once per cycle will maintain the amplitude of the motion.

A Class C amplifier is one in which the valve or transistor is biased well beyond cut-off and only the extreme peaks of one half-cycle of the input cause conduction. The angle of flow may be only 30 degrees, but during this period sufficient energy will be supplied to the tuned circuit to maintain the amplitude of the sinusoidal output voltage. The efficiency is very high, but the amplifier can only be used for one frequency—the resonant frequency of the tuned circuit.

A Class B amplifier is one in which the angle of current flow is 180 degrees, and the efficiency is thus greater than in Class A. The valve or transistor is biased just to cut-off, and one half-cycle causes current to flow.

The load is not a tuned circuit, but an undistorted output is obtained by using a pair of valves or transistors in what is known as a 'push–pull' arrangement. Each of the pair amplifies a different half-cycle, and the two outputs are combined in a single load resistor or transformer to give full sine waves. The advantage of extra efficiency is obtained while still retaining the Class A ability to amplify a wide range of frequencies.

If the application of suitable feedback is used to convert an amplifier into an oscillator, the general efficiency conditions outlined above will still apply.

Different conditions apply in these various amplifiers and oscillators, e.g. only in oscillators will there always be signal present and only in Class A is there always current in the active device. Because of this a variety of different methods are used to obtain the required bias conditions.

Relaxation Oscillators; Multivibrator

In the oscillators described above, a small fraction of the alternating voltage available at the output terminals is fed back to the valve or transistor input and sinusoidal oscillations are maintained at a frequency determined by the tuned circuit.

There is another important type of oscillator in which the whole of the voltage appearing across the output terminals is applied to the input of the transistor or valve. As a result of this massive feedback, the working point is taken not merely into the non-linear region of the characteristic but beyond it, so that the valve or transistor is actually cut off for comparatively long periods. These are sometimes called relaxation oscillators, and the output consists of almost square voltage pulses of a duration and frequency determined by resistors and capacitors in the circuit.

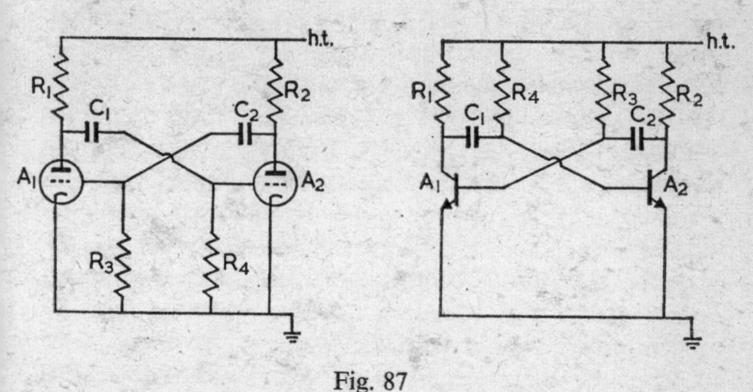

Fig. 87

Fig. 87 shows a relaxation oscillator called a multivibrator—valve and transistor versions.

Although the position of some of the components is shown slightly differently, the multivibrator is essentially a simple two-stage amplifier, with the output of the second stage connected back to the input of the first stage. The reader should confirm this by inspection of the two circuits.

When the multivibrator is working, the output, which can be taken from either of the two anodes or collectors, consists of a stream of square voltage pulses. The two active elements in the circuit are alternately cut off and fully conducting, with one cut off and the other conducting at any instant.

Let us consider in more detail a cycle of operation of the circuit, starting with the assumption that the left plate of C_2 suddenly falls in potential. We shall look at what then happens and see how the cycle proceeds to a point at which the left plate of C_2 is suddenly driven down in potential, i.e. the cycle restarts.

When the left plate of C_2 goes down in potential, so does the grid or base of A_1 and it is cut off. No current now flows through A_1 or the load R_1, so the anode or collector rises to the full h.t. potential.

The left plate of C_1 is connected to the bottom of R_1 and will also rise in potential. The right plate of C_1 will instantaneously rise by the same amount as the left, because it takes time for charge to flow in a circuit to change the potential difference between the plates of a condenser. This sudden rise in the potential of the right plate of C_1 will take the grid or base of A_2 up with it and ensure that A_2 is fully conducting.

Consider again the situation at the left plate of C_2, where the potential has gone down sharply, cutting off A_1. Current will now flow in R_3, tending to restore the left plate of C_2 to its original potential.

This rise in potential will take place exponentially at a speed determined by the time constant of the circuit $(C \times R)$. During this time A_1 remains cut off, but after an interval, largely determined by C_2R_3, the potential of the left plate, and hence the grid or base will reach a value at which A_1 conducts.

As A_1 conducts, current flows in R_1 and the potential at its bottom end falls. This fall is transmitted through C_1 to the grid or base of A_2, which is thus cut off. A_2 then remains cut off for a time, determined by the time for C_1 to charge through R_4 in the same way as was described above for C_2R_3.

When A_2 again conducts, the potential at the bottom end of R_2 falls sharply, and this fall in potential is transmitted through C_2. The left plate of C_2 falls sharply in potential, giving us the situation from which we started to describe the cycle.

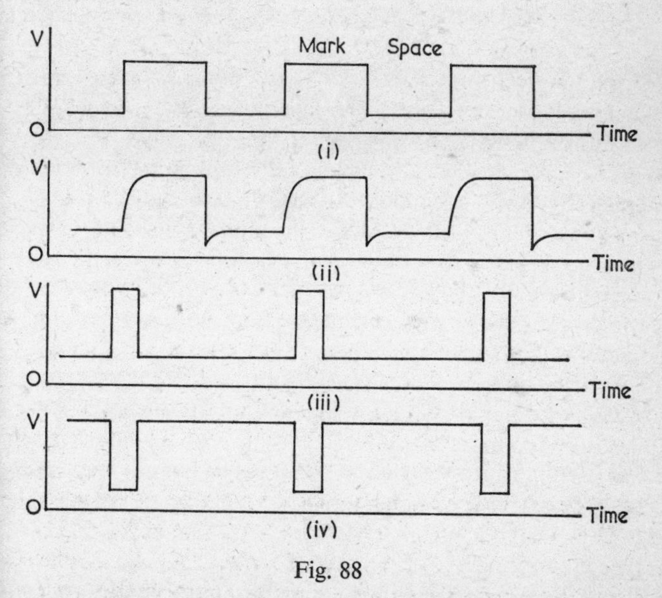

Fig. 88

The anode or collector is thus switched up and down between the h.t. and a much lower potential, and the output from the multivibrator, taken from one of the anodes or collectors, has the idealised form shown in Fig. 88 (i).

The amplifying action and the 100 per cent feedback are essential in producing the sudden switching of the active elements between the cut off and the fully conducting states. A small change in potential at one grid or base causes an amplified change in the opposite direction at the output of that element. This output voltage is fed back to the input of the other stage and further amplified with change of sign to give, at the original grid or base, a very much larger change of potential in the same direction as the small change which initiated the process.

In practice, due to additional capacity in the circuit and other effects, the 'square' waves will probably have rounded corners and spikes something like those shown in (ii). If necessary, the shape can be improved by various electronic methods, including clipping the top or bottom with diodes, as outlined on page 141.

The time each active element is cut off depends upon the appropriate *C* and *R*, and if the two halves of the circuit are symmetrical the mark/space ratio will be unity, i.e. each active element will be on and off for equal times.

If the two halves are not symmetrical, the on–off times will not be the same, and the output from one of the valves or transistors will be short pulses with long gaps (iii) and the output from the other will be complementary, with long pulses and short gaps (iv).

Monostable and Bistable Multivibrator

In the free-running, or astable, multivibrator, the steady bias of the two stages is such that, if they were quite separate, they would both be conducting. It is only the transmission from one stage to another of a sudden fall in potential which causes cut-off to occur. But this is not a stable state and if, for instance, A_1 is cut off it only remains so for the time taken for a capacitor to charge. A_1 then reverts to the conducting state. In doing so, however, it transmits a fall of potential to A_2, which cuts off and then takes a certain time to revert to conducting.

In the monostable multivibrator, a steady bias is applied to one of the active elements—let us say A_1—so that it is normally held beyond cut-off. The stable state for the circuit is thus with A_1 off and A_2 on.

If a pulse from some external source is applied to the input of A_1, then, providing the pulse is of sufficient amplitude and the correct polarity, A_1 will conduct. At this instant, A_1 will transmit a fall of potential to the input of A_2, which will cut off. The consequent rise in potential at the output of A_2 is fed back to the input of A_1, maintaining it in the conducting state.

But A_2 will only remain cut off for the time taken for the appropriate condenser to charge. When A_2 reverts to the conducting state, it will transmit a fall in potential to the input of A_1, which will cut off because by now the external pulse has ended and the fixed bias on A_1 is no longer opposed by a potential from the output of A_2.

Once A_1 is again cut off, it will remain so until the arrival of another suitable pulse from the external source.

The arrival of the pulse on the input of A_1 will cause the output terminal of A_1 to rise in potential, remain at that high potential for a certain time, determined by the appropriate CR value, and then fall again to the original potential. There will be one square pulse of fixed length and amplitude delivered by the circuit for every suitable pulse provided at the input. As long as the polarity of the input pulse is correct and its amplitude exceeds a certain value, there is little restriction on its shape.

The monostable multivibrator will give out a standard square pulse when triggered by an input pulse, which, within broad limits, can be of any shape.

In the bistable multivibrator, both active elements are

biased so that there are two possible steady states for the circuit: A_1-off, A_2-on, and A_1-on, A_2-off.

If A_1 is off, its output will be at a high potential and this will keep A_2 on, in spite of the bias. But if a suitable pulse is applied to A_1, the bias will be overcome, A_1 output potential will fall and so will A_2 input, thus cutting off A_2. A_2 will remain cut off because of the bias and A_1 will remain on. Only when a suitable pulse is now applied to the input of A_2 will A_2 bc switched on and A_1 be cut off.

If the circuit output is taken from A_2, it will be a square pulse, rising in voltage when the first trigger pulse arrives at A_1 input and lasting until the second trigger pulse arrives at A_2 input. The onset of the output pulse and its duration are controlled by trigger pulses from two different sources, one applied to A_1 and the other applied to A_2.

As it stands, the bistable multivibrator has applications in pulse generation, and suitably modified it has an important role as a counter (see page 176).

CHAPTER 9

Applications of Switching

The switch with which we are most familiar is the one in which we make a mechanical movement and thus turn off or on the electrical power to a lamp, an oven, a television receiver, or the motor in a refrigerator or washing machine. Sometimes the need for human action may be eliminated by incorporating a clockwork mechanism or a thermostat so that the switch operates unaided at a certain time or temperature. Occasionally, such automatic action may be electronic, e.g. a photoelectric cell may provide the input to operate a valve or transistor switch.

But the outstanding feature of an electronic switch is the very high speed with which it can be turned off and on. This is not very important in the above applications. It is predominantly in the field of electricity generally, and electronics particularly, that phenomena occur so quickly that high-speed switching is needed. Consequently, electronic switches are most often applied in these fields, where they provide the only relative simple solution to the problem of speed, and where their power requirements and connections, and properties generally, are compatible with the rest of the equipment. We shall consider below some typical applications of this type.

Rectification

For reasons associated with the economics of generation and transmission, the basic electrical power available in any community is usually an alternating supply.

The supply voltage fluctuates, usually sinusoidally, as

shown in Fig. 89, and the frequency is generally 50 or 60 Hz. This can be used unmodified for heating, lighting and some electric motors. But there are many applications where the load being supplied requires a potential difference of fixed polarity to be applied to it, i.e. it needs

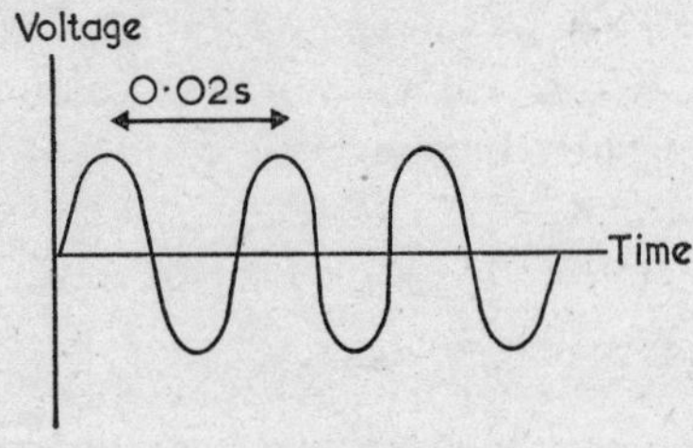

Fig. 89

a unidirectional current flow. Where such a d.c. supply is required, the mains a.c. supply is rectified using circuits like those described below.

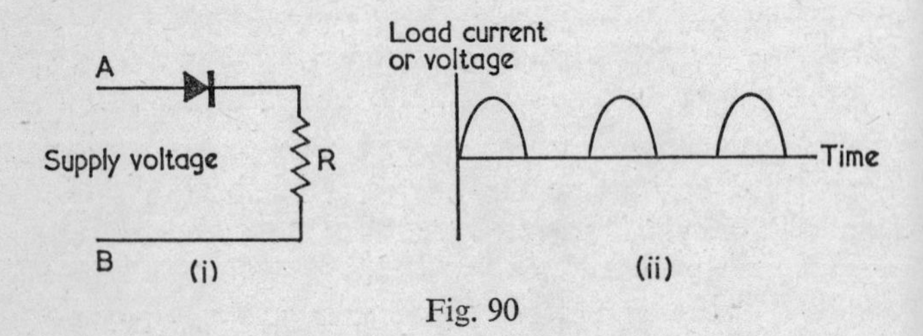

Fig. 90

In Fig. 90 (i) the sinusoidal supply voltage is connected between *A* and *B*. When *A* is positive with respect to *B*, the diode will conduct and current will flow through the load, which is represented by the resistor *R*. In the opposite half-cycle of the mains voltage, the diode will not conduct and no current will flow in the load.

The diode acts as a switch which only connects the load to the mains when the voltage is of the required polarity.

For a sinusoidal supply voltage, the current flowing in the load, and the potential difference across it, consists of half-sine waves, as shown in Fig. 90 (ii). Such a circuit is called a half-wave rectifier.

Although the output current of the rectifier is always in the same direction, it has a far from steady value and would be most unsuitable for many loads. Two steps can be taken to secure a smoother output from the rectifier.

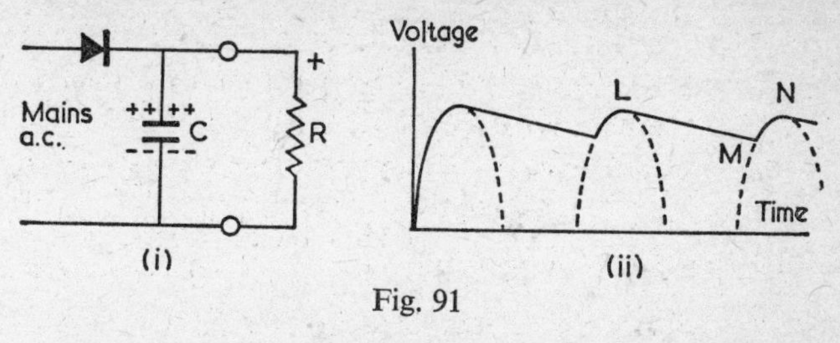

Fig. 91

In Fig. 91 (i) a reservoir condenser C is placed across the rectifier output terminals in parallel with the load. When the diode conducts, current flows through the load as before, but it now also charges the capacitor C, as shown. In the half-cycles, when the diode is not conducting, current will continue to flow through the load in the correct direction because the capacitor will now discharge through it.

The potential difference across the rectifier terminals is shown in Fig. 91 (ii), where LM indicates the gradual fall in voltage as the condenser discharges through the load during the non-conducting half-cycles. MN shows the voltage recovering during the conducting half-cycles.

The fluctuation in output voltage has now been reduced

to a comparatively small 'ripple'. This ripple can be regarded as an a.c. voltage superimposed on the required d.c. output, and it can be further reduced by incorporating an additional large capacitor and a large inductance, or choke, as a smoothing circuit.

In Fig. 92 such a smoothing circuit has been added to reduce the a.c. ripple. If a given a.c. potential difference exists between the points P and Q, this same potential difference will exist across the path PXY. Along this path, part of the voltage drop will occur across PX and

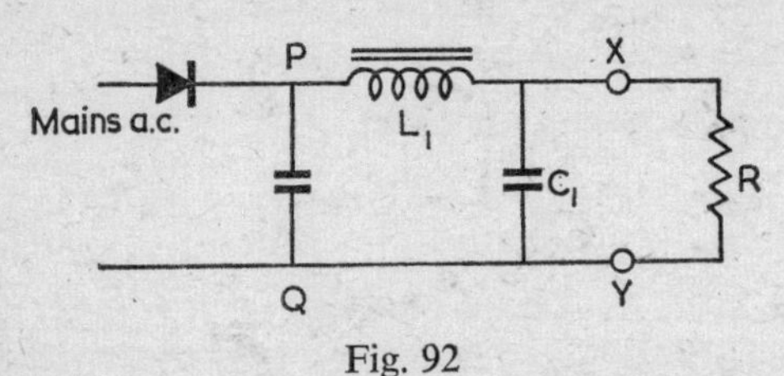

Fig. 92

the remainder across XY, in the ratio of the impedance of the inductor L_1 to that of the capacitor C_1. These components are chosen so that a large fraction of the a.c. voltage between P and Y is dropped between P and X, and only a very small fraction of the original ripple now appears across the output terminals XY.

The rectifier circuit shown in Fig. 93 (i) has the mains a.c. supplied to it from the secondary of a transformer, which has a centre tap at X.

In the half-cycle of the mains voltage, when A is positive with respect to X, the diode D_1 will conduct and the current will flow through R. In the other half-cycle, B will be positive with respect to X and the diode D_2 will conduct. Current will flow through the load in the same direction as when D_1 conducts. In the absence of the reservoir condenser and the smoothing circuit, the voltage appearing across the output terminals and the external load *R* would have the form shown in Fig. 93 (ii).

The diodes alternately switch the top end of *R* to the ends *A* and *B* as they become positive with respect to *X*. The voltage applied to *R* is thus that between *AX* or *BX*, i.e. half the transformer secondary voltage. The reservoir condenser and the smoothing circuit have the same effect

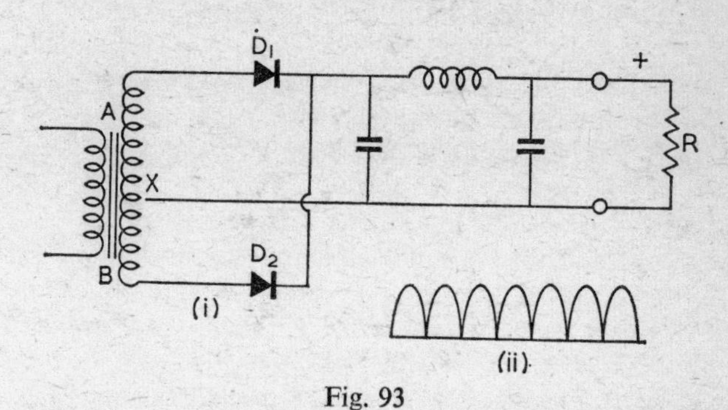

Fig. 93

as before in providing a steady d.c. output. This circuit is called a full-wave rectifier.

Demodulation

In Chapter 1 (page 24) we have seen that a common method of transmitting information, particularly speech and music, is by modulating the amplitude of a radio-frequency carrier wave.

The alternating current produced in the receiving aerial then has the form shown in Fig. 94, and it is the function of the radio receiver to process this weak signal so that the audio-frequency envelope is extracted from it and made available at a sufficiently high-power level to drive a loudspeaker or similar acoustic transducer.

If the modulated r.f. signal of Fig. 94 is applied to the terminals *AB* of the circuit in Fig. 95 (i), the diode will conduct during the positive half-cycles of the r.f. voltage, and current will only flow in the resistor *R* during these periods. If there were no capacitor in the circuit, the

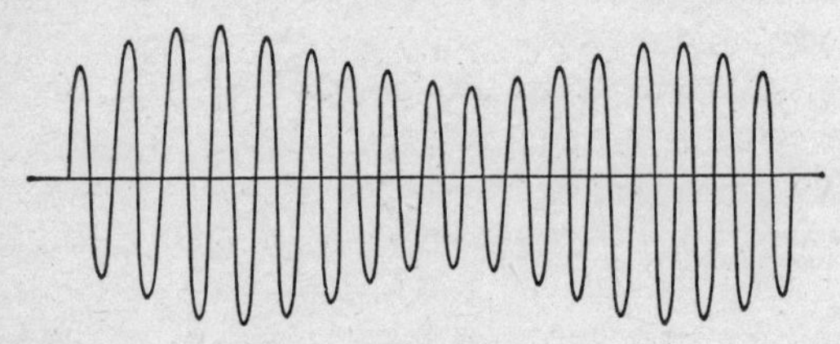

Fig. 94

current in *R*, and the voltage across it, would have the form shown in Fig. 95 (ii). But the addition of the capacitor *C* has the same effect as the reservoir capacity in the rectifier circuits described above.

The capacitor charges during the conducting half-cycles, and the peak voltage reached during one half-

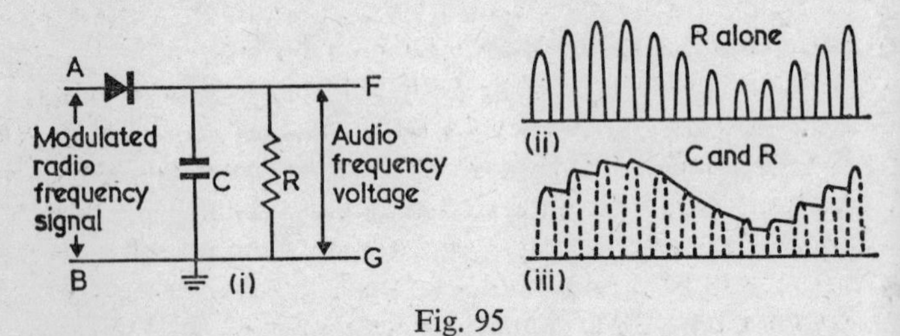

Fig. 95

cycle tends to be held until the next because discharge takes place only slowly through *R*. The voltage across *C*, and hence across *R*, now has the form of the full line in Fig. 95 (iii). The speed at which a capacitor charges or discharges through a resistor is determined by the time constant C (in farads) $\times$ R (in ohms). In this circuit, CR (in seconds) should be long compared with the periodic time of the r.f. signal and short compared with the periodic time of the a.f. modulation.

Under these circumstances, the output of this detector or demodulating circuit, which appears between F and G, is an audio-frequency voltage with a very slight r.f. ripple on it. This voltage will be passed to an audio-frequency amplifier and thence to the loudspeaker, which will respond to the a.f. and be unaffected by the r.f. ripple.

Gates, Logic and Computation

The reader will have met applications of ordinary mechanical switches in 'logic' circuits. Common examples are those for turning a hall light on or off from upstairs or downstairs, and for making a lift stop only at floors where people want to be picked up or dropped.

The switch shown in Fig. 96 allows the terminal C to be connected either to terminal A or to terminal B.

A B C

Fig. 96

Arrange two of these switches in circuit with a battery and a lamp so that the lamp may be turned on and off from either switch. A suitable circuit is shown in Fig. 97 overleaf and it is suggested that the reader now attempts to solve this problem, before looking at the diagram or reading further.

If the pair of wires between the two switches in the diagram are called 0 and 1 respectively, then the light is on when the switch positions are 00 or 11, but off when the positions are 01 or 10.

The block shown in Fig. 98 represents a reversing switch. With the switch in one position, the line labelled 0 is connected to the line P and 1 is connected to Q. With the switch in the other position, 0 is connected to Q and 1 to P.

Using the logical approach of the previous paragraph, where 00 and 11 for the end switch positions represented lamp on, it is comparatively easy to see that any number of such reversing switches may be inserted into the double line joining the end switches. The lamp may now be switched on and off from the two end switches, or from any of the interposed reversing switches.

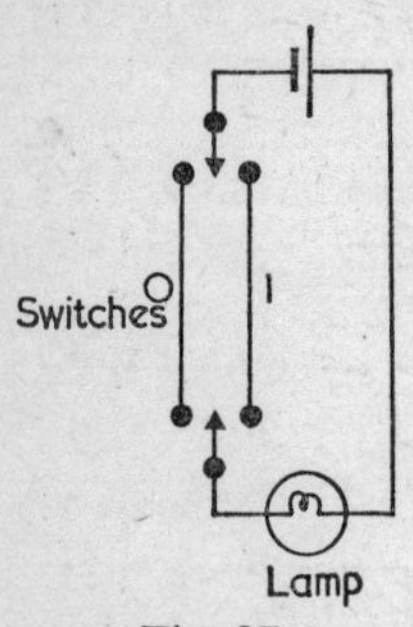

Fig. 97

The symbols 0 and 1 have been chosen arbitrarily, but the reader will be aware that these are the numerals used in the binary arithmetic of the modern electronic computer. In this field, more than any other, enormous numbers of electronic switches are used.

On page 142 we saw how two diodes could be used together to make a special switch called an AND-gate. In this section we shall consider how this gate and other similar switching circuits can be made to carry out some of the fundamental processes of computation.

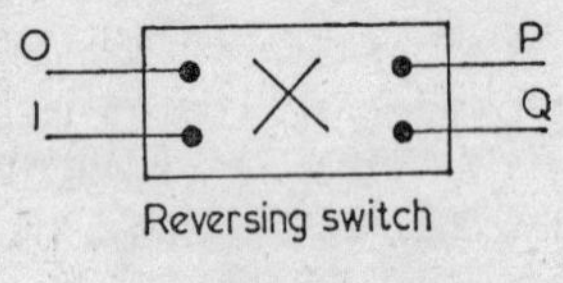

Fig. 98

Let us first make a list of some of the component parts used to carry out arithmetical operations.

AND-gate

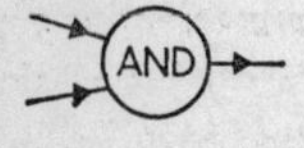

A pulse on both input lines is required to give a pulse on the output line.

OR-gate

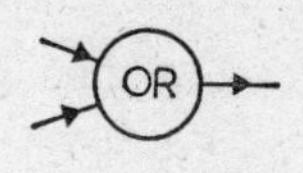

A pulse on either one or the other of the input lines gives a pulse on the output line. This is often referred to as an exclusive OR-gate because a pulse on *both* input lines gives zero output.

Negater

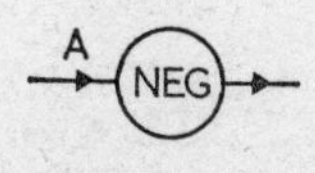

Whenever there is a pulse on the input line, there is zero output. Whenever there is zero input on line A, there is a pulse generated on the output line. The output is said to be the negation of the input.

A negater can be made by feeding clock pulses to one input of an OR-gate and the number that is to be negated to the other input.

Clock pulses are regularly spaced continuously generated pulses (see page 162), whose function is to provide the datum time scale on which the computer operates. Each clock pulse defines an instant of time and, at any point in the computer, the 1 or the 0 of a binary number is represented by the presence or absence of a pulse at that point at the instant the clock pulse occurs. Thus the number 11 would be represented by two pulses occurring at the instant of two successive clock pulses, while 101 would be represented by a pulse, a gap and a pulse respectively occurring at three successive clock pulses.

The reader should satisfy himself that, if an intermittent stream of pulses representing a binary number is applied to one input of an OR-gate and a continuous stream of clock pulses is applied to the other input, the output will be an intermittent stream of pulses which is the negation of the number fed in, i.e. all the 1's in the original number become 0's and all the 0's become 1's—1101001 becomes 0010110.

Before combining the above units into a simple binary adding circuit, it is worth recalling the rules of binary arithmetic.

Only two digits, 0 and 1, are used. The various digits in a number represent 2^0, 2^1, 2^2, 2^3, etc., just as in the decimal system the various digits represent 10^0, 10^1, 10^2, 10^3, etc.

Decimal: 347 is $7 \times 10^0 + 4 \times 10^1 + 3 \times 10^2$, i.e. three hundred and forty-seven.

Binary: 1101 is $1 \times 2^0 + 0 \times 2^1 + 1 \times 2^2 + 1 \times 2^3$, which is the binary equivalent of thirteen.

In binary addition $0 + 0 = 0$, $0 + 1 = 1$ and $1 + 1 = 0$, with 1 carried to the next place:

$$\begin{array}{r} 1101 \\ +1001 \\ \hline 10110 \\ \hline \end{array}$$

The arrangement shown in Fig. 99 below is for the addition of two binary digits, *A* and *B*. The result of this addition is two digits, *S* and *C*. *S* is the sum digit, which must be written in the same significant place as *A* and *B*. *C* is the carry digit, which must be added to the next

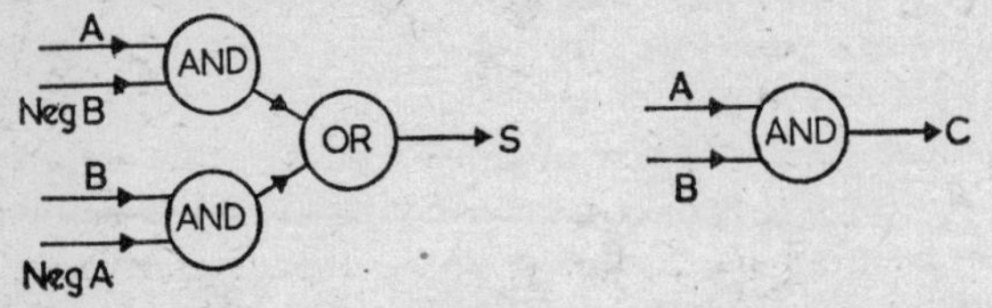

Fig. 99

highest significant place. The table shows the possibilities which may arise.

A	B	S	C
0	0	0	0
0	1	1	0
1	0	1	0
1	1	0	1

The first step in the addition of the digits A and B is to generate their negations, which is assumed to have already been done using negaters. Three AND-gates are now fed with A and the negation of B, B and the negation of A, and A and B, as shown.

The output of the right hand AND-gate, which is fed with A and B, is the carry digit C.

The outputs of the other two AND-gates are fed into an OR-gate, whose output is the sum digit S.

Take the first line of the table when $A = 0$ and $B = 0$.

Then neg. $A = 1$ and neg. $B = 1$.

Input to the top AND-gate is 0 and 1, so output is 0.

Input to bottom AND-gate is 0 and 1, so output is also 0.

The inputs to the OR-gate are thus both 0 and the output—the sum digit—is also 0.

The inputs to the right hand AND-gate are both 0, so the output—the carry digit—is also 0.

The reader should check the other three lines of the table in the same way to confirm that the arrangement shown does give the correct S and C in each case.

The arrangement described above allows the addition of two binary digits. The output is a sum digit to be written in one significant place and a carry digit to be added to the next significant place.

If we require to add two binary *numbers*, i.e. two collections of binary digits, then for each significant place there will, in general, be two digits from the two numbers to be added, plus the carry digit from the previous significant place. There will be three inputs to the adder, instead of two, and two outputs as before, with the output carry digit being handed on for addition to the next significant place.

There will be a more elaborate arrangement of individual components, but the general principles and the operation of the individual components will be as described in the simple adder above.

Counting Circuits

The switching applications described above have largely been based upon diodes, although the same functions could usually have been performed—occasionally more effectively—with triodes or transistors. In Chapter 8, under the applications of the amplification principle, relaxation oscillation has been described, and this could be regarded as a switching process, with the two active elements alternately switching each other on and off. In the monostable and bistable multivibrators, the switching can be considered to be carried out by the triggering pulse or pulses.

The bistable multivibrator gives out one pulse for every two triggering pulses put in. But the two input pulses must be on different lines, each connected to the input of a different one of the active elements.

This circuit can be modified so that the input pulses all come along one line and are then directed to the correct active element to make the switching action occur. Thus for every two pulses on the input line there is one pulse on the output line. This circuit is known as a binary counter, or scale of two circuit.

A bistable multivibrator modified with pulse steering diodes is represented in Fig. 100 below.

A_1 and A_2 are the two active elements, which are assumed to have a positive bias at the input when conducting and a negative bias when cut off.

The triggering pulses are of positive polarity and enter as shown along a single line connected to the junction of two diodes, each of which leads to the input of one of the

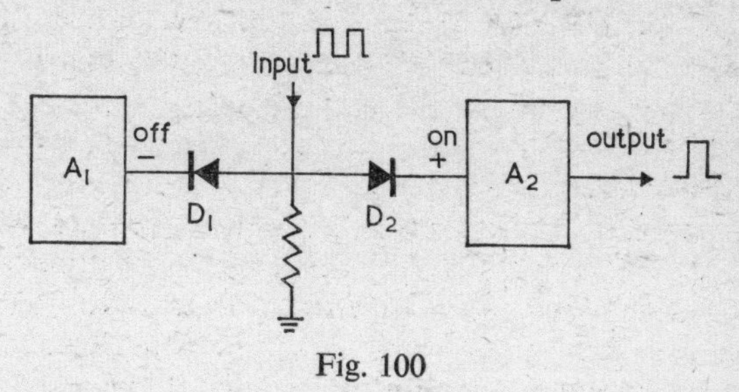

Fig. 100

active elements. The purpose of the steering diodes is to ensure that the incoming positive pulse is always directed to the input of the element which is cut off so that switching occurs.

In the above diagram A_1 is assumed to be off, so its input is negative in potential, while A_2 is conducting and will have a positive potential at its input. Consequently, D_1 will be biased in the forward direction and will be a short circuit, allowing the incoming positive pulse to reach A_1 and switch it. D_2 is biased in the reverse direction and thus is effectively an open circuit which will not allow the positive pulse to reach A_2.

When A_1 switches on, A_2 will switch off by the multivibrator action. Consequently, the next pulse to arrive will be directed to A_2, which will now switch off and the circuit will be back in its original state.

Two pulses have come into the circuit, and a single pulse will come out as the output of A_2 rises in potential when it cuts off and falls when it conducts once more.

Using a bistable circuit modified in the above way, or some similar circuit, a pair of active elements can be made to act as a binary, or scale of two, circuit. If the output

0	⊖	⊕	⊖	⊕	⊖	⊕	
1	⊕	⊖	⊖	⊕	⊖	⊕	
2	⊖	⊕	⊕	⊖	⊖	⊕	
3	⊕	⊖	⊕	⊖	⊖	⊕	
4	⊖	⊕	⊖	⊕	⊕	⊖	
5	⊕	⊖	⊖	⊕	⊕	⊖	
6	⊖	⊕	⊕	⊖	⊕	⊖	
7	⊕	⊖	⊕	⊖	⊕	⊖	
8	⊖	⊕	⊖	⊕	⊖	⊕	Final output pulse

Fig. 101

of such a circuit is fed to the input of an identical one, then a scale of four counter is produced. Three binary counters in series make a scale of eight counter, giving one pulse out for every eight pulses put in.

Fig. 101 above shows the state of affairs in a scale of eight counter as eight pulses are received and one is handed on. The three pairs of active elements are shown with + to indicate ON and − to indicate OFF.

The first line, 0, indicates the state of the circuit before the first pulse enters. The first, or left, element of each pair is cut off and the other is therefore conducting. The reader should trace the changes in the circuit shown on the following lines of Fig. 101 as the eight pulses pass through the counter.

For every eight pulses received, there will be one final output pulse from the last stage of the circuit. If N final output pulses have been delivered, then the total number of pulses received is $8N$ plus some number less than eight, which can be determined by examining the state of the circuit to see which of the lines in Fig. 101 it corresponds to. Some sort of pointer or lamp display normally allows this state to be seen easily.

Such electronic counting can be done extremely fast, and with many binary stages in series it is possible to scale down a large number of pulses received in a short time until they can be recorded by a mechanical counter which is comparatively slow acting. The extra significant figures in the answer are again given by the state of the circuit when the count is ended.

For direct use in computation or process control, the count can be left in binary form. However, when a counter output is to be directly read or printed, it is usually much more convenient to have it in decimal form.

A binary counter with four stages will operate in the scale of $2^4 = 16$, but it can be modified to the decimal form.

Normally, the circuit would accept fifteen pulses, and the sixteenth would then give a final output pulse and reset the circuit to zero, i.e. to the state it was in before any pulses arrived.

When modified, the first nine pulses act just as before, but the change in the circuit is so arranged that when the tenth pulse arrives it gives a single final output pulse and at the same time resets the circuit to the state it was in before any pulses arrived. Six such decade circuits will give a scale of one million counter.

CHAPTER 10

Electronic Systems: Communication, Computation and Instrumentation

In a general way, we have seen in Chapter 1 what electronics can be made to accomplish. The needs of modern society for systems of communication, control, computation, etc., were discussed and the principles of the techniques used to satisfy these needs were outlined. Elsewhere, we have looked at the minutiae of electronics: the individual components, valves and transistors, and the ways in which they work. On a slightly larger scale we saw how small numbers of these components—the clay and straw of electronics—could be put together as amplifiers, gates or counters to make the bricks from which the systems we require are built.

The systems chosen for examination here are communication, computation and general instrumentation. These are areas covering the most numerous, the most common and the fastest-growing types of electronic application.

Generally, we shall consider the systems as collections of 'black boxes'. In most cases, the boxes contain circuits already understood. In other cases, however, the standard circuits will be slightly modified. Occasionally, there will be an important feature to discuss which has not previously been met.

Communication Systems

In Fig. 102 is shown a straightforward system of radio communication suitable for speech and music.

The basis of the transmitter is the oscillator which

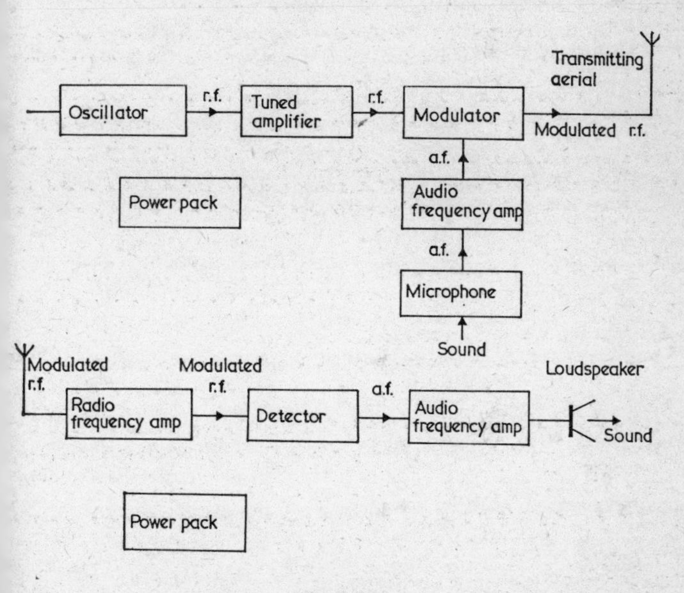

Fig. 102

generates the radio-frequency carrier wave. The frequency will be determined by a parallel tuned circuit and, if the oscillator output is not great enough, it will be amplified. Because the r.f. carrier consists of a single frequency, it may be amplified by a tuned amplifier. This carrier frequency is then applied to the modulator stage, which is discussed later.

The sound information which is to be placed on the r.f. carrier wave is converted by the microphone into a low frequency electrical signal and amplified in the audio-frequency amplifier. The frequency of the signals to be handled by this amplifier will, in general, cover the whole audio range, and all must be amplified by the same amount if faithful reproduction of speech and music is to occur.

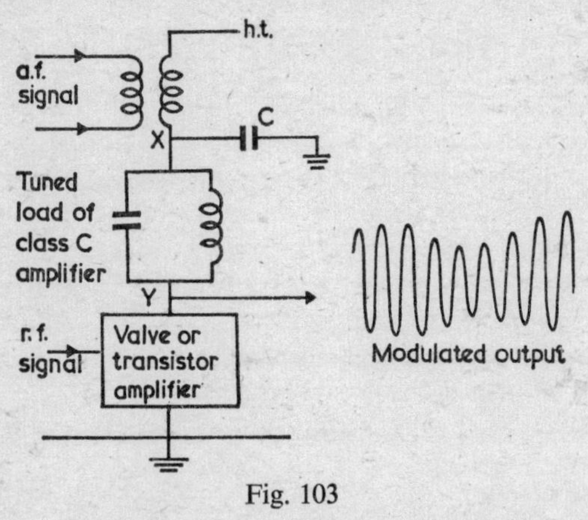

Fig. 103

The a.f. amplifier will normally have resistance load stages to give a large bandwidth.

The type of modulation used in this system is amplitude modulation, and it must now be arranged that the r.f. carrier amplitude fluctuates up and down in accordance with the audio-frequency signal which is to be transmitted. This process is carried out in the modulator, and it can be done in several ways. In the method described, the h.t. voltage supplied to a Class C tuned amplifier is caused to vary by the audio-frequency signal.

In Fig. 103 above is shown a valve or transistor amplifier, with a tuned circuit load resonant to the carrier frequency. If the point X were connected directly to the h.t., then it would be a simple tuned r.f. amplifier with bias (not shown) to make it operate in Class C. Under these circumstances, the output taken from Y would have been just the carrier frequency.

It is a property of a Class C amplifier that the gain depends upon the h.t. voltage, i.e. for a constant input the amplitude of the output is almost proportional to the h.t. voltage. Thus, if the point X is made to fluctuate up and down in potential, the amplitude of the r.f. output will fluctuate correspondingly.

In the circuit shown, the audio-frequency signal is used to make the potential of X fluctuate above and below the steady h.t. value. This is achieved by inserting the secondary of a transformer between X and the h.t. line, the primary being fed with the amplified a.f. signal.

The potential of X will now fluctuate in the required manner, and the output available at Y will be as shown; the r.f. carrier amplitude is modulated by the a.f. This output is fed to the aerial and a modulated electro-magnetic wave is radiated.

The capacitor C is of such a size that it represents a short circuit to earth at the frequency of the carrier, but a high impedance to the audio-frequency. Thus no r.f. current flows through the secondary of the transformer, and the only fluctuation in potential at X is at the audio-frequency.

Included in the transmitter system is a separate block labelled power pack. This may be a set of batteries, or it may take a.c. power from the mains and, with trans-formers and rectifiers, convert it into the d.c. supplies required for h.t. and bias, and, if valves are used, supply a low voltage output for the heaters.

In the receiver shown, the modulated r.f. signal is first

amplified and then applied to the detector or demodulator. The audio-frequency output of the detector passes through the a.f. amplifier and thence to the loudspeaker. As in the transmitter, there will also be a separate power supply unit providing h.t., etc., for all the stages.

Once the r.f. carrier has been modulated, each unit through which the signal subsequently passes must have sufficient bandwidth to accommodate the carrier and the sidebands. The same bandwidth criterion applies to the link between the transmitter and the receiver. In this case the link is free space, which has no intrinsic bandwidth limitation. But there must be no other system operating at a frequency close enough to interfere, so in this way the number of systems which can be accommodated by free space is limited.

Modulation of a carrier may still be used even when the signal is carried by cable or wire. By using different carrier frequencies, a number of separate messages can be sent simultaneously along the same cable, providing each carrier is arranged so that it and its associated sidebands do not interfere with other messages on neighbouring frequencies.

Improved performance is often achieved in radio systems by substituting a superheterodyne receiver for the straight receiver shown above.

In the superhet receiver shown in Fig. 104, the majority of the amplification before demodulation takes place at a fixed frequency called the intermediate frequency (i.f.).

The output of a special variable frequency oscillator in the receiver is mixed with the incoming r.f. signal in a mixer stage. This stage contains a diode or some other device with a non-linear I–V characteristic which enables a signal at the sum or difference of two incoming frequencies to be generated. Suppose we extract the difference frequency and pass it to the i.f. amplifier. Thus the radio-frequency minus the local oscillator frequency is equal to the intermediate frequency.

The tuning of the radio-frequency stage and the local oscillator are 'ganged' together so that they are both controlled by a single knob and the i.f. is always the same. The i.f. signal retains the modulation of the original r.f., and, after amplification in the i.f. amplifier, demodulation and a.f. amplification occur as in the straight receiver.

The communication system described above has been based upon amplitude modulation. There are many other forms of modulation, and these have all been developed as a result of the careful study of what is meant by information and what factors are involved in its transmission and reception.

We have already touched upon some of the factors in Chapter 1, where we saw that the more sophisticated the information to be transmitted, the greater the bandwidth that was required. We saw that the speed of transmission was significant in that a single picture could be sent slowly over a system of small bandwidth (facsimile process), but that the rapid transmission of many pictures per second, as in television, demanded much greater bandwidth. Electronic noise is always operating against the satisfactory trans-

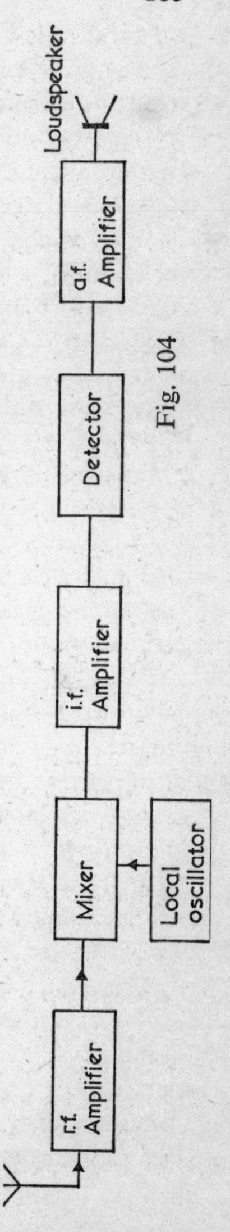

Fig. 104

mission of information. This may come from outside the system, due to disturbance in the medium between the transmitter and receiver, or it may come from inside the system, because every electronic stage handling the signal will to a small or large extent be a generator of noise.

We need not concern ourselves unduly with the precise way in which these factors interact in determining the capacity of a given system, i.e. the rate at which it can carry information, but the equation linking capacity and bandwidth is as follows:

$$C = B \log_2 (1 + S/N) \text{ bits per second.}$$

In this expression, B is the bandwidth, S/N is the ratio of the signal voltage to the noise voltage and C is the capacity of the system. The unit of information here is the bit, which is derived from the words 'binary digit', because of the special way in which information is defined in communication theory.

Instead of making the amplitude of the carrier change in accordance with the audio-frequency, it is possible to keep the amplitude constant and change the carrier frequency by an amount depending on the instantaneous amplitude of the audio-frequency it is required to transmit. Such a system is called frequency modulation and it has advantages over amplitude modulation with respect to noise.

There are a number of modulation systems which use pulses instead of a continuous carrier wave. The amplitude of the pulses may be modulated (PAM), or the amplitude may be kept constant and the pulse width modulated (PWM). One obvious advantage of these and similar pulse systems is that one or more other sets of pulses can be transmitted in the gaps between the pulses of the first set, thus achieving multi-channel operation over a single link.

There is a form of pulse modulation, increasingly used in modern systems, which comes close to the best performance that can be achieved according to the present state of communication theory. This is called pulse code modulation (PCM).

Noise means that the transmission of certain information about small changes in signal amplitude is wasted because these signal changes are less than those which might be produced by noise. Pulse code modulation recognises this and transmits no information of signal changes below a certain value.

There will be a limit to the signal amplitude, and the scale between zero and this amplitude will be divided into a number of steps—say, ten. The instantaneous amplitude of the signal will always lie within one of these steps.

Each step is identified by a particular set of pulses—a code—which is transmitted to indicate that the signal amplitude has that particular value. Thus a three-digit binary code, e.g. 101, could indicate one of a total of eight different levels.

Although these details of communication systems have been related principally to radio systems, many of the operations and principles are common to line communication, and the reader can expect to find them used in new systems such as communication by laser beam through space or a 'light pipe'.

Computation

Essentially in the digital computer, electronic switches, gates, etc., are arranged so that numbers can be added and subtracted, compared and transferred from one place to another. These are only a comparatively small number of operations, but if they are done many times, and in a carefully chosen sequence, they can be made to solve very difficult problems. It is the job of the programmer to work out a suitable sequence for each problem and then put it into a form—e.g. a punched tape—that can be accepted by the program unit.

In Fig. 105 the important sections of a digital computer are shown in block form, although the units performing

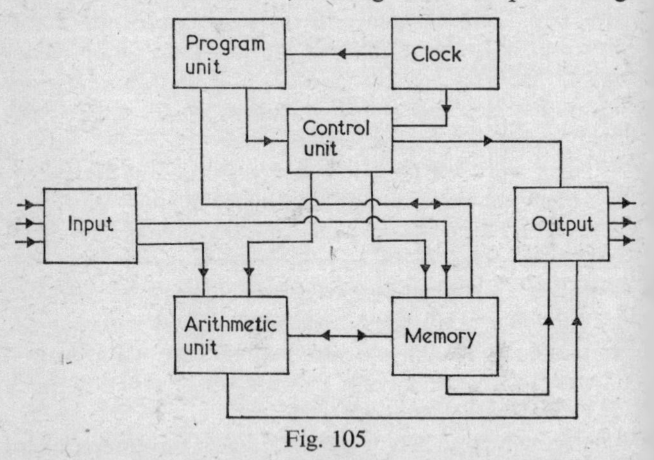

Fig. 105

one particular function may often be distributed throughout the machine and not confined to one box.

There may, for instance, be several memory units: some long-term stores, and others only holding for a short while the binary digits that make up all the information in the machine.

The long-term store may be a magnetic tape, disc or drum with facilities for rapid write-in and read-out of information. A more elaborate form of magnetic store with a much faster access time for information is made from a matrix of many tiny ferrite rings, each threaded by several wires. A *zero* is represented by magnetisation in one direction and a *one* by the opposite magnetisation. Pulses of current can be sent along the wires to write information into any particular ferrite core and also to interrogate a core, i.e. to discover the direction of magnetisation. Another wire brings out the information that a 0 or a 1 was stored.

The capacity of a permanent store will usually be very large, but often temporary stores are required which retain a single piece of information for a short period. Usually this information will be a set of twenty or thirty binary digits called a word. Such a store—sometimes called a register—is often made of a set of bistable multivibrators, with zeros and ones indicated by the conducting state of the multivibrators at any instant.

The computation processes are carried out in the arithmetic unit by circuits like those described in Chapter 9. It is important that the numbers required in these processes are taken from the input or the memory at the right time and in the right order, and made available at the right point in the arithmetic unit.

The program should contain the correct sequence of operations, and the basic timing of these operations is fixed by the clock, which may be a crystal oscillator with the output arranged to provide regular spaced voltage pulses giving the basic rhythm of the machine. The program unit and the clock together operate the control unit, which arranges that all the required numbers are presented to the arithmetic unit at the correct time and that the answer, when available, is supplied to the output or the memory.

In the block diagram a number of leads are shown into the input unit and out of the output unit. The input unit must accept information from various sources and make it available in a form acceptable to the machine. Typically, these sources might be punched tape, punched cards, a manual keyboard, the output of another computer, or measurements in digital form from the instruments in a process control system. Similarly, the output unit may be required to feed a typewriter for a printed output, a tape or card punch, a digital display, or the control equipment of some process.

All these machines connected to the input and output units are sometimes referred to as the peripheral equipment, while the rest of the units are the computer hardware. The term 'software' is often used to describe the programs and routines produced externally by human programmers. Now that digital computers have existed long enough for their immediate advantages in pure speed and accuracy to have been realised, it is being increasingly recognised that expertise and inventiveness in the software field are the main requirements for future progress.

Instrumentation

Electronic computation and communication involve some extremely large, expensive, exciting and highly publicised systems and pieces of equipment. But there are some very important applications on a much smaller scale, involving as a rule some sort of transducer and an associated amplifier and display. Such instrument systems are in wide use in industry, medicine and research for the detection and measurement of all kinds of phenomena.

The instruments used for various measurements differ principally in the transducer. This is the device at the input which converts the quantity to be measured into an electrical signal. Thus the input to the instrument may be from, say, a thermocouple, a photoelectric cell, a geiger tube, or a piezo-electric crystal. The temperature, light, nuclear radiation, or the sound respectively are converted into voltages, which are fed into equipment differing only by details in the amplifiers and the display.

The thermocouple consists of two different metals or semiconductors, usually in the form of wire, connected as shown in Fig. 106 so that there are two junctions between the two different materials. If one junction is at

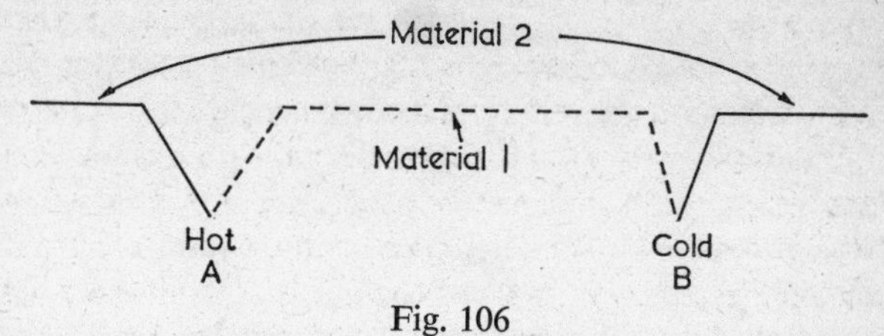

Fig. 106

a higher temperature than the other, then a potential difference appears between *A* and *B*. Usually one junction is kept at a fixed low temperature, and the potential difference between *A* and *B* varies with the temperature of the hot junction. The relation between voltage and temperature is not linear, and the instrument must be calibrated. The thermoelectric voltages generated will be in the microvolt range and will be amplified before measurement. In most cases, the variation in temperature will be very slow by electronic standards and consequently the amplifiers used must have a good response at the lowest frequencies. They will probably, therefore, be direct-coupled, although sometimes this sort of voltage is switched on and off by a 'chopper' switch, and the resultant voltage pulses amplified by an ordinary amplifier.

The geiger tube takes advantage of the fact that

nuclear radiation, α- or β-particles or pulses of γ-rays, has sufficient energy to ionise gas molecules. The gas is contained in a tube with a thin glass or metal 'window' to admit the less penetrating radiation if this is to be measured. Inside the tube is a cylindrical metal electrode, with another electrode consisting of a wire running down the central axis of the cylinder, as shown in Fig. 107.

The voltage maintained between the two electrodes is, by itself, not sufficiently great to cause the gas in the tube to break down and conduct. When a pulse of radiation appears, ions are produced in the gas which move at high speed to the appropriate electrodes, causing further ionisation by collision on the way. A pulse of current flows round the circuit, and the resultant voltage pulse across R is taken away for amplification and counting. The electrode geometry is such that the electrical discharge quickly dies out and the tube is ready to receive the next pulse of radiation.

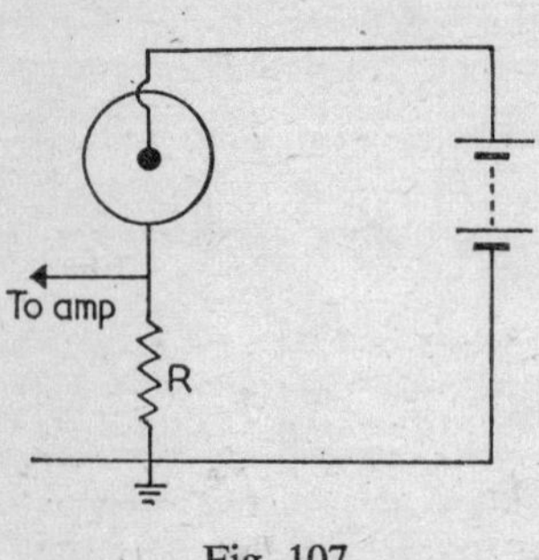

Fig. 107

In the photoelectric cell, there is a cathode with a low work function from which light will cause electrons to be emitted. When this occurs, a current flows to a nearby anode—just as in the diode valve—and a voltage appears across the load resistor, which is in series with the cell and the h.t. supply. This voltage is then amplified in the usual way before measurement, or before being used to operate a relay controlling some other circuit.

The piezo-electric crystal converts the mechanical vibrations due to a sound wave into an alternating voltage and vice versa. The piezo-electric materials, e.g. quartz or barium titanate, possess the property that, if pressure is applied between two opposite faces of a rectangular slice, a potential difference will appear between two other faces. If the pressure is alternating, as with an incident sound wave, then an alternating voltage is produced. Used in this fashion, the crystal can act as a microphone converting sound energy into electrical signals which can be amplified before use, e.g. to modulate an r.f. carrier or to work a remote loudspeaker in a public address system.

Operated in the converse way an alternating voltage applied to the crystal will produce sound waves of the same frequency. Since the crystal operates at audio- and much higher frequencies, it is commonly used to generate ultrasonic waves. If only a single frequency is to be generated or detected, then the crystal is cut to a particular size to give mechanical resonance at that frequency and thus increase the efficiency of the transducer action.

Ultrasonic energy may be generated for a number of simple uses, e.g. agitating liquid mixtures to give emulsions, but some of the more interesting applications involve the generation of an ultrasonic pulse which is projected into a region where objects or inhomogeneities can be detected and mapped by the reflections of the pulse. Generally, there is a transmitter and a receiver, and the method is used in echo sounding and submarine detection, in flaw detection in large castings and in medical diagnostics.

These are some of the ways in which conventional electronics is exploited in instrumentation. In the next chapter are discussed some unconventional electronics, such as masers and travelling wave amplifiers, and some ways of making up conventional circuits by new techniques.

CHAPTER 11

Further Electronic Developments

With such a large and rapidly growing subject as electronics, a short book must omit much detail. The fundamental principles and techniques have already been covered, and in this chapter it is intended to give a brief, self-contained account of some of the more important innovations that the reader is likely to meet.

Masers and Lasers

This is perhaps the largest single field of electronic development in the last few years, with a wide range of equipment now available, and enormous capital investment in current production and new research.

The underlying principle—stimulated emission of radiation—was first discussed by Einstein as long ago as 1917, but was not successfully applied until 1954. The result is an amplifier operating on quantum principles.

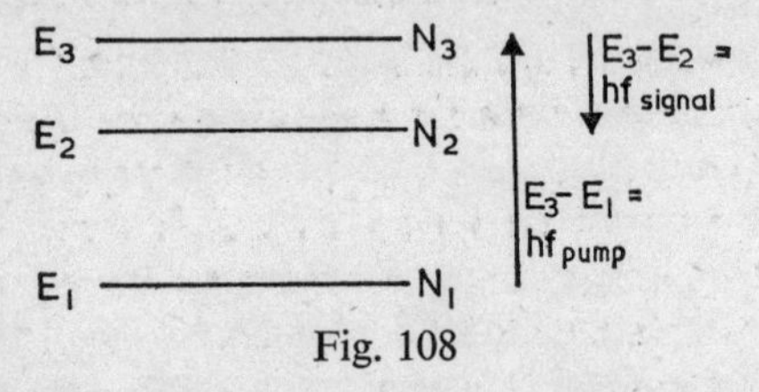

Fig. 108

Suppose that we have a material containing units—probably atoms—with three energy levels $E_3 > E_2 > E_1$, as shown in Fig. 108.

Let us first concentrate on levels E_3 and E_2, where the frequency of a signal we wish to amplify is f_{signal} and

$$E_3 - E_2 = hf_{\text{signal}}.$$

In the natural state of normal equilibrium in a system, there will always be more units in the lower of two permitted energy levels than in the higher. The number of units in a given energy state is called the population N, and for the two levels E_3 and E_2, which are our immediate concern, $N_2 > N_3$ in the equilibrium condition.

Suppose we consider signal energy entering the material which is in the equilibrium state. The signal will consist of many quanta, each of energy $hf_{\text{signal}} = E_3 - E_2$. Thus, if one of these quanta meets an atom in the lower energy state E_2, it will be absorbed and will raise the atom to the higher state E_3. This is the normal selective absorption process and it will cause attenuation of the signal.

The stimulated emission of radiation is the converse process and takes place when a quantum of signal energy is incident on an atom in the higher energy state E_3. The effect of the incident quantum is to stimulate the atom to fall from state E_3 to E_2 and emit a further quantum of energy which is identical in every respect—including frequency, phase and direction—with the incident quantum. Thus, where there was originally one quantum of signal energy, there are now two and the signal is amplified.

Some of the signal quanta are lost by absorption and some are doubled by stimulated emission. But under normal conditions there are far more atoms in the material in the low energy state than in the high one. Consequently, there are far more absorptions than stimulated emissions, and the net result of passing a signal through a normal material is that attenuation of the signal strength occurs.

If we could arrange that there were more atoms in the high state than the low one, there would be more stimulated emissions than absorptions and amplification of the signal would be the net result. The required state, when $N_3 > N_2$, is called population inversion. It is a non-equilibrium state and energy must be supplied to the material to achieve it. However, if energy were supplied to achieve population inversion by raising atoms from E_2 to E_3, this energy would be at the same frequency as the signal to be amplified and confusion would result.

Accordingly, the population N_3 is increased by raising to E_3 atoms from a much lower energy level, E_1. The source which supplies this energy to the amplifier is called the pump, and it is analogous in purpose to the h.t. supply in an ordinary amplifier. The pump frequency is f_{pump}, where $E_3 - E_1 = hf_{pump}$. This is quite different from the signal frequency, and the material can be continuously pumped without any danger of interference to the signals being amplified.

The choice of the active material for a quantum amplifier depends principally upon the existence within it of energy levels suitable for the system described above. Frequencies from the microwave to the optical region are handled by different amplifiers, and the active materials used may be solids or gases. Secondary considerations in the choice of active material are the ease with which it can be pumped, the convenience with which it can be used, and the sharpness of the energy levels.

If atoms with the required energy levels are used in the gaseous form, then the concentration of active material is low and, to get the required amplification of a signal, it must have a very long path in the gas. Gaseous active materials are mainly used in lasers—light sources—and a neon laser may be a metre long. On the other hand, a greater concentration of active atoms may be achieved by having them dispersed as impurities, say 1 per cent, in an inert solid. Ruby contains such a small concentration of chromium, and in a synthetic ruby laser the length of active solid material may be only a few centimetres. But the output light from a solid laser lacks the spectral purity of that from the gas laser because the greater concentration of atoms in the solid broadens the sharp energy levels.

In general, gas lasers are likely to be used in applications where a stable, pure frequency is more important than the output power, e.g. in holography where a three-dimensional image of an object can be re-created from a two-dimensional photograph. The photograph, or hologram, is not recognisable to the eye as a picture of the object, and it is, in fact, a complex pattern due to the interference of two beams of light from the same laser: one falling directly on the photographic film, and the other reaching the film after being scattered by the object. Once the pattern is recorded on film it may be used to produce a three-dimensional image which is almost indistinguishable from the original object. In re-creating the image, light identical with that which originally produced the hologram must be used, and this is obtained from a laser similar to that used in exposing the film. Once again, two beams of light are needed: one is passed through the hologram, and the other, direct from the laser, interferes with it to produce a perfect image of the object.

A simple view of the process is that the hologram was originally produced by light from the object (Ob) interfering with light direct from the laser (La), so that the hologram may be represented as La + Ob. In reconstructing the image, the light passing through the hologram is La + Ob, and if this is made to interfere with direct light (La) in a suitable fashion, then the resultant light may be considered to be (La + Ob) − (La) = Ob, i.e. a pattern of light is established which is exactly the same as that which came originally from the object, and the effect upon the eye will be as if the object were actually there. It will, for instance, be possible to look around a protrusion on the image at a hidden region by moving the eye to a different view point—something which cannot be done with an ordinary photograph.

The maser (*M*icrowave *A*mplifier by *S*timulated *E*mission of *R*adiation) is a low-noise amplifier used in the GHz region in high-quality communication, radar and radio astronomy systems, where the minimum noise is essential if weak signals are to be detected.

The pump energy is provided by an oscillator in the receiver generating an output at a considerably higher frequency than the signal to be amplified. This pump is required to maintain the non-equilibrium state, and it is opposed by the thermal vibration in the material which allows the constituent atoms to interact with each other. Such interaction tends to bring the material into the normal equilibrium state with larger numbers of atoms in the lower energy states. To reduce this interaction to a minimum the thermal vibration is made as small as is conveniently possible by cooling the maser to 4K in liquid helium.

In the maser a special method is used to achieve the required energy level system. The two levels E_3 and E_2 are produced at exactly the right separation by the application of a magnetic field, which causes a level existing originally in the material to divide into two new levels, one higher and one lower. The separation between the two levels increases with the strength of the magnetic field, which can thus be used to tune the amplifier to exactly the signal frequency.

The signal is fed into one end of a short piece of wave-guide, which contains a number of small pieces of ruby. In addition to the signal wave, energy from the pump is passed down the guide to cause population inversion in the ruby. The signal passes over the ruby causing stimulated emission and an amplified signal emerges from the far end of the guide. A magnetic field and cooling is supplied, and the whole system is known as a travelling wave maser.

The maser will usually be found mounted on the aerial, as in the ground stations for satellite communication systems. Low-noise amplification of the signal thus occurs before it is sent down the long feeder from the aerial to the main receiver. The maser unit will typically be self-contained, with its own liquid helium cooling system and a superconducting magnet. Such a receiving system will be several hundred times less noisy than a conventional microwave receiver and will thus be able to detect correspondingly weaker signals.

The laser, in spite of the name (*L*ight *A*mplification by *S*timulated *E*mission of *R*adiation), is really an oscillator, i.e. a light source. But we have already seen that an oscillator will, in general, consist of an amplifier with some arrangement for feeding back part of the amplified signal. We shall find that in the laser there is amplification by stimulated emission from a pumped active material, with feedback arranged using mirrors.

In Fig. 109 suppose that *A* is a block of active material which has been pumped into the non-equilibrium state. The two energy levels between which the stimulated emission transition occurs are so far apart that the frequency of the radiation is in the optical band.

There are two plane parallel mirrors at the ends of the active material, so a quantum of light—a photon—emitted in a direction perpendicular to the mirrors will tend to travel up and down through the active material in the manner shown by the arrows.

Such a photon will have a very long path in the active material and, if a population inversion has been achieved by pumping, there will be a very good chance of stimulated emission, and hence amplification occurring. A photon emitted in any other direction will not be reflected back along its original path and will quickly pass out of the active material without significant amplification.

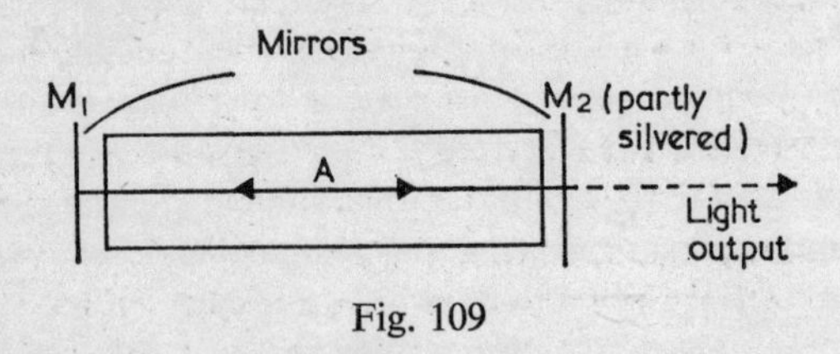

Fig. 109

At the instant of switching on the laser there are always a few photons spontaneously emitted in random directions by atoms in the higher energy state. Radiation of the 'correct' kind is continuously enhanced and the remainder dies out.

One of the mirrors is only partially reflecting, so that a fraction of the light emerges through it while sufficient is fed back by reflection to maintain the oscillation.

All the light from the laser is parallel and in the same phase, and because of this it can be focused to an extremely small spot so that the energy intensity is sufficient to drill holes in solids and weld substances together with great precision. Furthermore, since the light has this coherence, it can be subjected to all the complex modulation processes of modern communication technology and, because of the high frequency, the bandwidth available when techniques have been perfected will accommodate an enormous number of telephone, television, digital data, etc., channels.

The active material in a laser may be a solid, such as a synthetic ruby rod, or a gas, such as neon or carbon dioxide, in a glass tube. Solid lasers are usually pumped by supplying them with optical energy from a flash tube, while an electric discharge provides the pump energy in the gas laser, often with a second gas such as helium added to help in the transfer of energy from the charged particles to the active gas by collision.

Electron Beam Tubes

There is a range of special valves, largely used in the microwave region, which depend upon the interaction between a beam of electrons and some sort of unusual electric circuit. The kinetic energy of the electrons in the beam is converted into radio-frequency energy in the circuit, and in this way amplification or oscillation is achieved. Such valves depend critically upon the electron velocity and, in particular, upon the time they take to travel from one part of the valve to another. This transit time is one of the principles upon which such unconventional valves depend, whereas in conventional triodes, pentodes and transistors the effects of transit time severely limit the performance at the highest frequencies.

The klystron amplifier shown in Fig. 110 uses two resonant cavities as tuned microwave circuits. The cavities, shown in section, are closed circular metal 'boxes' with electron currents circulating at GHz frequencies in the inner surface of the walls, just as current circulates between the inductor and capacitor in a conventional tuned circuit. In the conventional circuit, energy is contained in the electric field associated with the charged capacitor and in the magnetic field associated with the current in the inductor. Similarly, there is energy in the alternating electromagnetic field contained within the resonant cavity. The resonant frequency of the cavity is determined by its internal size and geometry, which can be altered by screwing in a metal tuning slug.

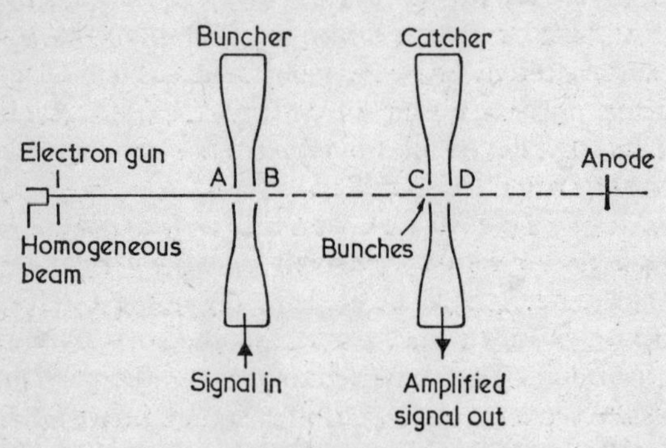

Fig. 110

A narrow beam of electrons flows from an electron gun to the anode, passing through holes in the centre of the two cavities. The wall of the cavity at *A* will be positive with respect to *B* during one half-cycle of the wave it contains and negative with respect to *B* during the other half-cycle. Thus half the electrons passing through the cavity are accelerated by the electric field between *A* and *B*, and the others are retarded. The amount of acceleration or retardation will depend upon the point in the r.f. cycle at which an electron crosses the cavity.

On leaving the cavity, the faster-moving electrons begin to overhaul those ahead of them which have been retarded until, after a certain distance, the original uniform beam is broken up into bunches of electrons and gaps between them. This first cavity is called the buncher.

If an electron is retarded by the electric field between *C* and *D* when passing through the second cavity, it will lose kinetic energy, which will go to increase the strength of the r.f. field which caused the retardation. Conversely, an accelerated electron would reduce the strength of the r.f. field in the cavity. This second cavity is called the catcher, and there is a continuous build up of energy in it because the bunches of electrons pass through it in the retarding phase and give energy to the r.f. field, while the gaps pass through it in the unfavourable phase.

Amplification is achieved by putting the signal into the buncher and by taking an amplified version of it out of the catcher. Very large klystron amplifiers are available which can deliver pulses with peak power in the megawatt range.

In principle, the amplifier could be converted into an oscillator by providing a feedback cable between the catcher and buncher. However, in practice, the feedback is effected through the beam itself, and the klystron oscillator uses a single cavity which acts as buncher and catcher. The beam passes through it into a region beyond called the drift space, where the electrons are gradually slowed down and turned back upon their original path by a negative potential on a reflector electrode. During its time in the drift space, the beam becomes bunched, and the bunches deliver power to the cavity when they pass back through it. This type of oscillator is called a reflex klystron.

In the klystron, the interaction between the electron

beam and the circuit containing the r.f. wave takes place at a well-defined point in space—at the lips of the cavity. Other electron beam devices exist where there is continuous interaction between the beam and the wave-containing circuit.

In the travelling wave tube amplifier shown in Fig. 111,

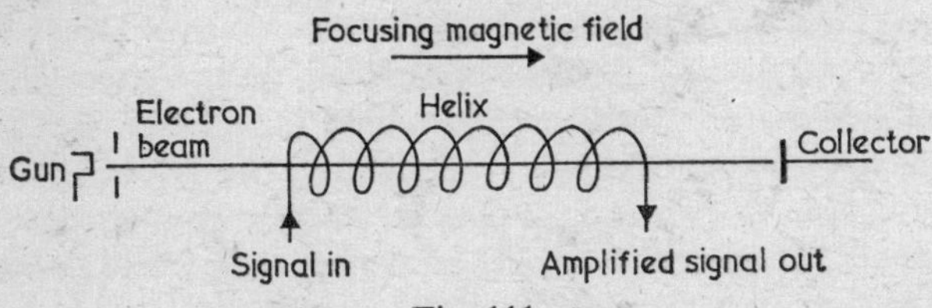

Fig. 111

the r.f. signal is carried by the helix, and its associated electromagnetic wave moves from left to right with a low effective velocity because it progresses as though travelling along the total length of the wire in the helix. The electron beam from the gun travels down the centre of the helix to the collector and is kept focused, so that it never touches the helix, by the magnetic field from a solenoid (not shown) which surrounds the whole length of the tube. The electron beam is arranged so as to have a similar velocity to that of the wave on the helix circuit, and there is a continuous loss of kinetic energy from the beam to the electromagnetic wave. The amplitude of the signal increases as it travels along the helix and an amplified version is removed at the far end.

The travelling wave tube is normally used as an amplifier in the GHz range and has the advantage of a large bandwidth, making it suitable for handling a signal which has been modulated by many communication channels.

There are other versions of electron beam tubes using complex circuits different from the helix to carry the r.f. wave. One of these, the backward wave oscillator, has an output whose frequency can be controlled within very wide limits by the voltage used to provide the electron beam velocity.

In the amplifying and oscillating devices above, the basic mechanism is the conversion of the kinetic energy of the charged particles into r.f. energy in a resonant cavity or other circuit. The converse process is possible where interaction between a beam of particles and an electromagnetic wave causes the particles to move faster and the r.f. energy to decrease. This is used to accelerate charged particles for nuclear research, the beam of particles being passed down a waveguide which is supplied with r.f. energy. This is called a linear accelerator and may be many hundreds of metres long.

Cryoelectronics

We have already met liquid helium as a necessary coolant in the maser, and in many electronic circuits cooling improves performance. Recent advances in the techniques of cryogenics—the achievement of very low temperatures—have meant that liquid helium, which boils under atmospheric pressure at 4K, is now more readily available to provide a low temperature environment outside the research laboratory. In particular, the phenomenon of superconduction is now being exploited industrially in the fields of heavy electrical engineering and electronics by using liquid helium to provide the necessary very low working temperatures.

If the temperature of a metallic conductor is gradually lowered, then the resistivity falls continuously. But there are some metals and alloys in which there is an extra effect: at a certain low temperature the resistivity of these materials disappears completely. Such materials are called superconductors, and the temperature at which they pass from the normal resistive state to the superconducting state is called the transition temperature. Lead, mercury, tantalum, aluminium and many metallic combinations are the commonest superconductors, and each has its own transition temperature at which its resistance becomes zero. Transition temperatures range from just above 0 K to just above 20 K.

The phenomenon of zero resistance occurs because, in the relatively few superconductor materials, it is possible for electrons to exist in a special state known as a Cooper pair. A Cooper pair consists of two electrons which are bound together as a result of the complex interaction of all the free and fixed charged bodies in the crystal structure. The two electrons move as a pair through the material and a certain amount of energy—the binding energy of the pair—must be provided if it is to be broken up into two independently acting electrons.

Current flows in an ordinary conductor when electrons acquire a drift velocity from the applied electric field in addition to their ordinary kinetic energy. Resistance is due to the scattering of electrons out of their drift direction by collisions. But, before the electrons in a Cooper pair can be so scattered, sufficient energy must be available to break the bond between them. At extremely low temperatures there is insufficient thermal energy in the material to break up the Cooper pairs, which therefore cannot suffer scattering and will thus drift unimpeded through the material when an electric field is applied, i.e. they experience no resistance.

There are therefore two requirements for superconduction to occur: the material must be one of those in which

Cooper pairs can exist, and the temperature must be so low that the Cooper pairs are not broken up.

At 0K all the electrons in a superconductor are in Cooper pairs, while above the transition temperature all the electrons are normal. At temperatures between the two, there will be some normal and some Cooper electrons present, but the material will still show zero resistance since the finite resistance of the normal electrons has the zero resistance of the Cooper electrons in parallel with it.

The material will pass from the superconducting to the normal state if its temperature exceeds the transition temperature. There are two other critical phenomena which will cause the material to show normal resistance: if the ambient magnetic field exceeds a certain value, and if the current density in the material exceeds a certain value. These three critical values will be different for different materials, and for any one material they will be interrelated, so that the actual transition temperature for a given material will be lowered if a magnetic field is applied. The figure generally quoted for the transition temperature is that for zero magnetic field.

The extinction of superconduction by an applied magnetic field is used in the cryotron switch, where a superconducting wire is switched from the zero resistance to the normal state by the magnetic field due to current flowing in another superconductor.

In Fig. 112 the tantalum wire is switched between the superconducting and normal states by the magnetic field due to current which may be passed through the niobium solenoid. The niobium always remains superconducting because it has a higher critical temperature and magnetic field than the tantalum, and the current flowing in it thus dissipates no energy.

If a current is induced in a superconducting ring, then,

as long as the material remains superconducting, the current will continue to flow undiminished, without any battery or other source of e.m.f. present. Such persistent currents have been observed flowing without loss for several years.

The persistent current flowing in a superconducting material can be used as the basis of an information store in a digital computer, and the cryotron switch can be used as the basic element for the computer logic. In either case, the power dissipation is extremely small because a large amount of the total current flow will be resistanceless.

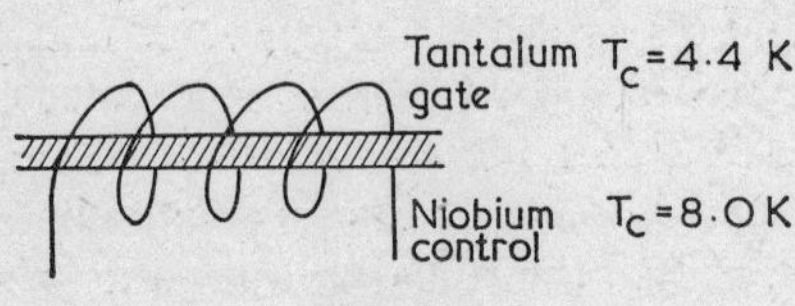

Fig. 112

These techniques are likely to be applied in the next generation of very large computers, and the many individual cryotron switches or persistent current cells needed will be obtained by depositing thin films of superconductors, suitably insulated from each other, in sheets containing thousands of elements.

There are applications, realised and proposed, for superconductors in the field of heavy electrical engineering, which really lies outside the scope of this book. But the superconducting motors, generators, power cables, electromagnets, etc., make use of a special superconductor property of general interest, and some of these heavy engineering applications will impinge on electronics, e.g. the use of a superconducting electromagnet with a maser.

In most of these applications, large current densities and large magnetic fields are encountered which might well cause ordinary bulk superconductors to pass to the normal state. But if a superconductor is prepared in thin film form, or if a thick wire is made from a superconducting alloy which has a filamentary structure, the critical magnetic field and current density will be higher than in a simple bulk superconductor. Superconductors of filamentary structure with high critical magnetic fields are called Group II superconductors and such materials, e.g. niobium-titanium, are used in heavy engineering applications rather than the simple Group I materials.

Other Solid State Devices

The three sections above have dealt with important subjects outside the mainstream electronics described in the rest of this book. But there are many devices which have developed directly from ordinary solid state diodes or transistors and which have become the core of important secondary electronic techniques. Some of these are mentioned below.

The nature of the junction between a *p*-type and an *n*-type semiconductor has been examined (see page 66) because of its important rectifying action. Two other *p–n* junction applications—the avalanche, or Zener, diode and the varactor diode—arise directly from the exploitation of the behaviour of the reverse current and the depletion layer respectively.

In Fig. 113 (i) *OA* is part of the forward current–voltage characteristic of a *p–n* junction diode and *OB* is that portion of the reverse characteristic which we have previously met. If the reverse voltage across the diode is further increased to V_c, the electric field across the depletion layer will become very large, and the energy acquired by the intrinsic charge carriers crossing the layer as reverse current will become sufficiently great to cause ionisation by collision, thus producing more charge carriers within the depletion layer. The charge carriers so produced add to the reverse current and will themselves acquire enough energy to cause further ionisation, and an avalanche of current will ensue. In addition to this avalanche current, the electric field may be great enough to excite electrons directly across the gap from the filled to the vacant band. This is called the Zener effect and further contributes to the large reverse current occurring at voltage V_c. At this point, the characteristic is ideally vertical (*CD*) and the diode may be used to maintain a stabilised voltage V_c across a load (see Fig. 113 (ii)), in spite of fluctuations in the supply or variations in the load itself.

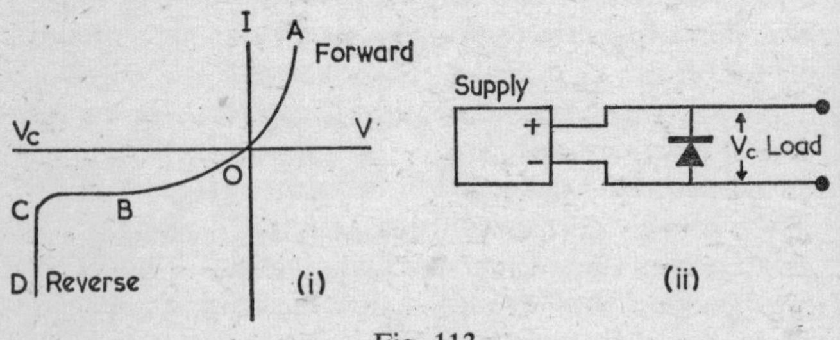

Fig. 113

The varactor diode, or variable capacitor diode, exploits the behaviour of the depletion layer which forms between the *p*-type and the *n*-type material at the junction. Within the depletion layer, there is charge stored in the form of fixed impurity ions no longer neutralised by the mobile charge carriers which exist in the bulk of the *p* or *n* material. The junction thus acts as a capacitor and the voltage applied to it will affect the size of the depletion layer, and hence its capacitance.

The varactor diode is normally biased in the reverse direction so that the 'leakage current' through the capacitor is very small, and any required variation in capacitance is obtained by varying this bias. Placed across a tuned circuit it allows the resonant frequency to be electrically controlled for remote tuning, automatic frequency control and frequency modulation.

A special application of the varactor diode is in a particular type of low-noise amplifier called a parametric amplifier. Here, the diode is placed in series with a load and the signal source. If the capacitance of the diode is now caused to vary, the amount of power in the load at the signal frequency may increase, i.e. amplification is achieved. The capacitance variation is brought about by a voltage source called the pump, which operates at a frequency above the signal frequency and is the prime source of amplifier power, as in the maser.

Another *p–n* diode application, which occurs in the crystal lamp and the injection laser, involves the emission of light from the junction.

When the *p–n* junction is biased in the forward direction, the current is made up of holes and electrons flowing in opposite directions. When these holes and electrons meet in the depletion layer, they recombine, with the emission of an amount of energy equal to the width of the forbidden gap. In germanium and silicon, the gap is comparatively small and only thermal energy is released. But gallium arsenide has a large gap and emits infra-red radiation, while gallium phosphide, with a still larger gap, gives visible light. The range of colours available can be extended by associating the infra-red emitter with phosphors, which give visible light when exposed to such radiation. These crystal lamps are used singly and in matrices for optical displays in electronic systems. The efficiency of the electricity–light conversion process is very high and, being essentially junction diodes, the lamps can usually be included directly in the circuit, which will be operating at appropriate current and voltage levels.

If two opposite faces of the light-emitting junction are made suitably plane parallel so that multiple internal reflection of emitted photons can occur, then, when the current is increased above a certain value, laser action can be obtained. Such an injection laser normally has the junction refrigerated and it can be operated continuously, with the light output modulated when required by correspondingly modulating the diode current.

Another important type of *p–n* junction is the tunnel diode, where the concentration of impurity in the two materials is extremely high, and a special state of affairs exists at the junction which encourages the presence of an extra component of current in the forward direction called tunnel current. Because of the wave-like nature of the electron, it is possible for it to penetrate a thin insulating layer from one material to another in which there is a suitable vacant energy level. In the tunnel diode this is possible for a limited range of voltages in the forward direction, and the resultant tunnel current added to the ordinary forward current gives an $I–V$ characteristic which has a negative slope over a certain voltage range, i.e. I increases when V decreases. Anything with such a characteristic can be made to produce amplification in a suitable circuit. Tunnel diode amplifiers operate at very high frequencies; they are used in high-speed pulse and switching circuits.

The diodes described above perform special functions other than simple one-way conduction. But there is a silicon rectifier now commonly used which has four layers, *p–n–p–n*, and three terminals. As in the ordinary diode, the terminal connected to the end *p* layer must be positive with respect to the terminal connected to the end *n* layer if conduction is to occur. But in this particular device, which is called a silicon controlled rectifier (SCR), or thyristor, there is a third terminal called the gate connected to the inner *p* layer. Even if the voltage applied between the main end terminals is of the correct polarity, conduction will not take place unless the gate is given a trigger pulse positive with respect to the terminal connected to the outside *n* layer. The gate is used to control the moment of conduction in big industrial rectifiers and switching circuits. However, once it has fired the conduction, it has no further control and current will continue to flow until the voltage between the main electrodes is reduced to zero or reversed.

Since the transistor was first invented in 1948, many different manufacturing methods have been used to improve performance and reliability, a particularly successful deposition technique resulting in the metal oxide silicon transistor (MOST). The majority of the great selection of transistors now available, covering a wide range of power and frequency, are of the conventional type with emitter, base and collector. There is, however, a special type called the field effect transistor (FET) where the corresponding electrodes are called source, gate and drain, and where the current flowing between source and drain depends upon the voltage applied to the gate rather than upon the current supplied by it. Because of this the FET acts much more like a valve than a transistor, where the current flowing between emitter and collector is controlled by the current in the base lead. The gate is equivalent to the control grid and, in particular, the FET has the very high input resistance associated with valves rather than the low value typical of the transistor.

Planar Epitaxial Transistors; Integrated Circuits

The most far-reaching electronic development to come from improved manufacturing techniques over the last dozen years is the process leading to the very accurately constructed planar epitaxial transistor and the integrated circuit.

First a thin slice from a single crystal rod of silicon of the required impurity concentration, *n* or *p*, is prepared and polished. Then an extremely thin layer of silicon is deposited from vapour onto the slice and oxidised so that a protective skin of insulating silicon dioxide covers the whole slice. The surface is then completely coated with a photosensitive material.

A tiny opaque mask with carefully positioned transparent holes is prepared by drawing it on a large scale and then reducing it photographically. This mask is placed on the slice and illuminated with ultra-violet light to expose the photosensitive material under the holes. The exposed parts are now dissolved away by a special photographic developer and the remaining unexposed material is converted by a baking process into a form impervious to the etching fluid in which the slice is next immersed.

The etching fluid dissolves the silicon dioxide except where it is protected by the unexposed photosensitive material. Finally, the unexposed photosensitive material is dissolved away and the slice is now left covered with an insulating layer except for the holes, through to the original doped silicon, corresponding to the holes on the mask.

It is now possible to make *p–n* junctions by depositing

onto the exposed material more silicon, but with impurity of the opposite 'polarity', *p* or *n*, to the original doped slice. By using different masks and repeating the process second junctions can be added to make transistors, thin layers of material can be used to make resistors and capacitors, and metal connections can also be laid down.

The general method described can be used to make single transistors, each on a silicon chip about $\frac{1}{10}$ mm^2 or integrated circuits each on a chip something over 1 mm^2 in area. But the original slice of silicon, a few square centimetres in area, has a lattice of squares, like graph paper, scratched upon it, each square being a single chip. In this way many hundreds of identical transistors or integrated circuits are made simultaneously on the single slice, which is then broken up to separate the individual chips. Each chip is then placed in a protective capsule which carries suitable connecting leads.

Instead of working with individual active and passive components the designer can now take whole circuits, like amplifiers, adders and counters, and assemble these into larger pieces of equipment with the functions he requires. The manufacturing processes, improving all the time, are now giving good yields of reliable integrated circuits with densities like 500 components in an area of 2 mm^2, so that quite complex units are available to be built up into the compact, multicomponent electronic systems of the future.

Appendix

The photographs give some impression of the development of electronic components and circuits over the last few years. In Plate 1 is shown a typical range of transistors. The larger ones with heavy metal cases have higher power ratings and are sometimes found attached to a heat sink—often the metal chassis on which the circuit is mounted—in order to increase still further their ability to dissipate heat, and hence to increase correspondingly the useful power which they can handle. In Plate 2, on approximately the same scale, is an integrated circuit showing the protective capsule and connecting pins. Such circuits have many components—transistors, diodes, resistors, etc.—all formed, with the appropriate connections between them, on a single minute silicon chip (see page 215). A magnified picture of such a circuit is shown in Plate 3.

The circuits in Plate 4 illustrate clearly the progress which has been made in miniaturisation. In the lower half of the picture is a decade counter (page 179) in which the separate components, including the transistors, may be easily identified. The output of the counter is applied to a display tube on which the number appears. This tube is the same size as a typical modern thermionic valve and it is thus possible to judge the saving in size which has been effected by using transistors rather than valves in the remainder of the circuit.

In the upper part of the photograph the same counter has been produced using integrated circuits. There is now a spectacular saving in space over the lower circuit. It is worth noting, however, that no comparable saving is possible in the display which has to provide a number big enough to be seen by the operator. A similar situation arises with a radio receiver, where much of the electronics could be reduced to integrated circuits but the loud-speaker must still occupy a large volume. Nevertheless, integrated circuits are likely to be used in such circum-stances because of their cheapness and reliability.

Index